Robert Jennings

The Horse and His Diseases

Embracing His History and Varieties, Breeding and Management and Vices

Robert Jennings

The Horse and His Diseases
Embracing His History and Varieties, Breeding and Management and Vices

ISBN/EAN: 9783337143787

Printed in Europe, USA, Canada, Australia, Japan

Cover: Foto ©berggeist007 / pixelio.de

More available books at **www.hansebooks.com**

THE

HORSE AND HIS DISEASES:

EMBRACING

HIS HISTORY AND VARIETIES, BREEDING AND MANAGEMENT
AND VICES; WITH THE DISEASES TO WHICH HE
IS SUBJECT, AND THE REMEDIES BEST
ADAPTED TO THEIR CURE.

By ROBERT JENNINGS, V.S.,

PROFESSOR OF PATHOLOGY AND OPERATIVE SURGERY IN THE VETERINARY COLLEGE OF
PHILADELPHIA; PROFESSOR OF VETERINARY MEDICINE IN THE LATE AGRI-
CULTURAL COLLEGE OF OHIO; SECRETARY OF THE AMERICAN
VETERINARY ASSOCIATION OF PHILADELPHIA, ETC., ETC.

TO WHICH ARE ADDED, RAREY'S METHOD OF TAMING HORSES, AND THE
LAW OF WARRANTY AS APPLICABLE TO THE PURCHASE
AND SALE OF THE ANIMAL.

Illustrated by nearly One Hundred Engravings.

PHILADELPHIA:
PUBLISHED BY JOHN E. POTTER,
NO. 617 SANSOM STREET.
1863.

PREFACE.

This volume is offered to the consideration of the public, not without a knowledge on the part of the author that many excellent works upon the horse have already appeared. It has occurred to him, however, that each of these various works is devoted rather to the consideration of some particular topic of interest in connection with this generous animal, than to a general treatment of the various subjects which appropriately claim notice in a work intended for the ordinary reader.

There are comparatively few in our country who are not, at some period or other, brought into contact with the horse, either as owners, hirers, or in some other capacity. The great majority cannot avail themselves of the numerous treatises already extant, which touch upon this animal, without gathering about themselves a library so large as seriously to trench upon their pecuniary comfort. Besides, so far as the ailments of the horse are concerned, much progress has been made in their treatment within the last few years. Old theories have been exploded, and, as the author believes, an era of a more humane and judicious medical treatment is dawning upon us. A marked improvement is discernible in the class of practitioners who essay the veterinary art; a greater amount of intelligence characterizes their action, and, as a consequence, the occupation of the veterinary surgeon is fast rising in popular estimation.

To these cheering indications of a better day for the horse and his owner, the author claims not to be indifferent. If an experience of fifteen years, diligently devoted to an investigation of the habits, peculiarities, wants, and weaknesses of the horse, has not been utterly fruitless in results, he flatters himself that he can at least contribute his mite in furtherance of a better understanding of an animal, which can never be too well understood by any one who would gain the greatest possible advantage from such a servant.

With such views the author has prepared the present work. Its pages are believed to contain a complete, candid, and truthful exposition of all the points which it is incumbent upon the horseowner to comprehend. Standard authorities upon the subject have been freely consulted, and the suggestions therein contained have been adopted, when corroborated by the author's own experience or observation. Reference has been made to the following among others:—Percival, Blain, Morton, Clark, Finley Dunn, Youatt, Coleman, and Spooner, on the Horse; Herbert's Horse of America, and Hints to Horsekeepers; Stewart's Stable Economy; The Farmer's Encyclopedia; and the Morgan Horse by Linsley.

The remedies recommended have all stood the test of actual trial, and are known to have proved efficacious in previous cases. As the author has no special hobby to ride, he has in this department of the subject given such modes of treatment only as he personally has superintended in actual practice, no matter from what source they may have been suggested. The very many illustrations throughout the volume it is believed will materially enhance its interest and value.

With the hope that the work may meet the approval of the large class for whom it was specially prepared, and with the consciousness that no effort to that end has been omitted by the author, he confidently leaves it in their hands, to be dealt with as to them shall seem most meet and proper.

Philadelphia, September, 1860.

CONTENTS.

	PAGE
HISTORY OF THE HORSE,	17
HORSES OF ASIA AND AFRICA,	24
The Arabian,	24
The Persian,	25
The Tartarian,	25
The Turkoman,	25
The Turkish Horse,	25
Horses of Hindostan,	26
The Barb and others,	26
EUROPEAN HORSES,	27
The Racer,	27
The Hunter,	27
The Hackney,	27
The Cart Horse,	28
GERMAN, FRENCH, AND SPANISH HORSES,	29
THE AMERICAN HORSE,	30
The American Blood Horse,	39
The Vermont Draught Horse,	57
The Conestoga Horse,	60
The Canadian Horse,	62
The Indian Pony,	64
The Narragansett Pacer,	65
The Morgan Horse,	69
NATURAL HISTORY OF THE HORSE,	73
BREEDING AND MANAGEMENT,	102
BREEDING,	102
BREAKING,	110
CASTRATION,	113
DOCKING,	120

BREEDING AND MANAGEMENT—Continued.

	PAGE
Nicking,	121
The Stable,	125
Air,	127
Litter,	130
Light,	131
Grooming,	133
Exercise,	135
Food,	138
Water,	153
Pasturing,	156
Service,	168
Shoeing,	175
Administering Medicine,	193

VICES OF THE HORSE, 198

Restiveness,	199
Baulking or Jibbing,	199
Biting,	203
Kicking,	204
Rearing,	206
Running Away,	207
Vicious to Clean,	207
Vicious to Shoe,	208
Crib-biting,	210
Wind-sucking,	212
Overreaching,	212
Not Lying Down,	213
Shying,	214
Pawing,	216
Rolling,	217
Slipping the Collar or Halter,	217
Stumbling,	218
Unsteadiness while being Mounted,	219

DISEASES, AND THEIR REMEDIES, 220

Diseases of the Mouth,	221
Lampas,	221
Inflamed Gums,	222
Bags or Washes,	222
Ulcers in the Mouth,	222
Sore Mouth,	223
Cut Tongue,	224

CONTENTS.

DISEASES, AND THEIR REMEDIES—Continued.   PAGE

- Uneven Teeth, ... 224
- Quidding, ... 225
- Wolf Teeth, ... 226
- Caries of the Teeth, ... 227
- Extracting Teeth, ... 230

DISEASES OF THE RESPIRATORY ORGANS, ... 231
- Inflammation, ... 231
- Sore Throat, ... 232
- Strangles, ... 234
- Chronic Cough, ... 235
- Catarrh, ... 236
- Distemper, ... 237
- Influenza, ... 237
- Bronchitis, ... 240
- Nasal Gleet, ... 241
- Pneumonia, ... 243
- Pleurisy, ... 245
- Hydrothorax, ... 247
- Thick Wind, ... 248
- Roaring and Whistling, ... 249
- Broken Wind, ... 249

DISEASES OF THE STOMACH AND INTESTINES, ... 251
- Inflammation of the Stomach, ... 251
- Inflammation of the Bowels, ... 252
- Diarrhœa, ... 255
- Inordinate Appetite, ... 256
- Palsy of the Stomach, ... 257
- Rupture of the Stomach, ... 258
- Calculus, or Stony Concretions, ... 259
- Hair Ball, ... 260
- Strangulation of the Intestines, ... 260
- Spasmodic Colic, ... 261
- Flatulent Colic, ... 262
- Worms, ... 263
- Bots, ... 265

DISEASES OF THE LIVER, ... 268
- Inflammation of the Liver, ... 269
- Jaundice, ... 270
- Hepatirrhœa, ... 271
- Decayed Structure of the Liver, ... 272

DISEASES OF THE URINARY ORGANS, ... 273
- Inflammation of the Bladder, ... 273

DISEASES, AND THEIR REMEDIES—CONTINUED.

	PAGE
Retention of Urine,	274
Profuse Staling,	275
Bloody Urine,	276
Stones in the Kidneys,	277
Stones in the Bladder,	278

DISEASES OF THE FEET AND LEGS, ... 279

Contraction of the Hoof,	279
Corns,	280
Quitter,	282
Thrush,	282
Canker,	283
Scratches,	284
Grease Heels,	284
Water Farcy,	286
Weed,	287
Cracked Hoof,	287
Sole Bruise and Gravel,	288
Pricking,	288
False Quarter,	289
Founder,	290
Pumiced Foot,	291
Corinitis,	291
Naviculararthritis,	292
Ossification of the Lateral Cartilages,	293
Wind Galls,	294
Sprung or Broken Knees,	294
Breaking Down,	295
Strains of the Knees,	295
Strain of the Hip Joint,	296
Shoulder Strain,	296
Open Joints,	297
Sweenie,	297
Ostitis,	298
Capulet and Capped Hock,	299
Caries of the Bones,	299
Bone Spavin,	300
Ring Bone,	303
Splint,	303
Curb,	304
String Halt,	305
Blood Spavin, Bog Spavin, and Thoroughpin,	305
Fractures,	306

DISEASES AND THEIR REMEDIES—Continued.

	PAGE
DISEASES OF THE HEART,	307
Pericarditis,	308
Carditis,	310
Endrocarditis,	310
DISEASES OF THE HEAD,	311
Osteo-Sarcoma,	311
Inflammation of the Brain,	312
Megrims,	313
Vertigo,	314
Epilepsy,	314
Stomach Staggers,	315
DISEASES OF THE EYE,	317
Amaurosis,	317
Inflammation of the Membrana Nictitans,	318
Simple Ophthalmia,	319
Specific Ophthalmia,	320
Cataract,	321
Wall Eye,	322
MISCELLANEOUS DISEASES,	322
Poll Evil,	322
Fistula of the Withers,	324
Melanotic Tumors,	324
Glanders,	325
Farcy,	327
Scarlet Fever,	328
Mange,	329
Surfeit,	331
Hide Bound,	332
Strains of the Loins,	332
Palsy,	333
Locked Jaw,	333
Rheumatism,	335
Cramp,	336
Hydrocele,	336
Warts,	337
Sit-Fasts,	327
Warbles,	338
Saddle or Harness Galls,	338
Mallanders and Sellenders,	338
Ulceration of the Udder,	339
Inflamed Veins,	339
SURGICAL CASES,	339

CONTENTS.

DISEASES AND THEIR REMEDIES—Continued.

	PAGE
Bleeding,	341
Neurotomy or Nerving,	342
Lithotomy,	345
Trephining,	345
Tenotomy,	346
Couching,	346
Tapping the Chest,	346
Periosteotomy,	347
Amputation of the Penis,	347
Œsophagotomy,	348
Hernia,	349
Roweling,	350
Firing,	351
Tracheotomy,	352

RAREY'S METHOD OF TAMING HORSES, ... 353

How to Call a Colt from Pasture,	356
How to Stable a Colt without Trouble,	357
Approaching a Colt,	362
How to Halter and Lead a Colt,	364
How to Tie up a Colt,	367
How to Tame a Horse,	368
How to Make a Horse Lie Down,	369
To Accustom a Horse to strange Sounds and Sights,	370
To Accustom a Horse to a Drum,	371
To Teach a Horse to bear an Umbrella,	372
To Fire off a Horse's Back,	372
How to Accustom a Horse to a Bit,	372
The proper Way to Bit a Colt,	373
How to Saddle a Colt,	374
How to Mount the Colt,	376
How to Ride a Colt,	378
How to Break a Horse to Harness,	379

WARRANTY, ... 382

ILLUSTRATIONS.

	PAGE
The Arab and his Steed,	17
The Shetland Pony.—An English Sporting Scene,	21
The Stallion,	28
American Farm Scene,	30
The Canadian,	35
Black Hawk,—An American Racer,	43
American Plantation Scene,	47
The Vermont Draught-Horse,	58
A Conestoga.—The Great Pennsylvania Draught-Horse,	60
Ethan Allen,—A Fast-trotting Morgan Horse,	69
Skeleton of the Horse, as covered by the Muscles,	73
Names applied to the various External Parts of the Horse,	80
Eight Days' Teeth,	88
Three or Four Months' Teeth,	88
Teeth at Twelve Months,	90
A Grinder sawed across,	90
Two Years' Teeth,	91
Three Years' Teeth,	92
Four Years' Teeth,	94
Five Years' Teeth,	95
Six Years' Teeth,	96
Seven Years' Teeth,	97
Eight or Nine Years' Teeth,	98
Mare and Foal,	102
The Arab Stallion, Jupiter,	105
Breaking,	110
The Agriculturist's Method,	117

ILLUSTRATIONS.

	PAGE
The Usual Method,	123
The French Method,	124
Customary Forms of Stalls,	126
Grooming,	133
Exercise,	136
Out to Grass,	143
The American Racer, Black Maria,	147
Pasturing,	157
Service,	169
Ground Surface of the Hoof,	175
The Hoof of the Horse,	176
A Section of the Foot,	177
The Position of the Shoe,	187
The Proper Form of a Shoe,	192
Running Away,	198
Particularly Dangerous,	204
Muzzle for a Crib-Biter,	210
Disagreeable and Dangerous,	214
The Sick Horse,	220
The Blooded Mare Fashion, and Foal,	227
The Saddle-Horse,	232
Quiet Enjoyment,	237
The Trotting-Horse, Lexington,	244
The Attack and Defense,	250
Omar Pasha, the Turkish Chieftain,	256
Sir Archy, the Godolphin of America,	262
Common Gad-Fly or Bot,	265
Eggs on a Hair,	266
Eggs Magnified,	266
Caterpillar, full size,	267
Caterpillar or Larvæ, adhering to the Lining of the Stomach,	267
The Red Gad-Fly,	268
Caterpillar of the Red Gad-Fly,	268
Virginia Mill-Boys on a Race,	273
The Fast-Trotting Stallion, Geo. M. Patchen,	279

	PAGE
The Children's Pet,	285
The Famous Trotting-Mare, Flora Temple,	292
The Equestrienne,	298
The High-Bred Pacing Mare, Pocahontas,	302
The end of Pericarditis,	309
Haying Scene,	316
The Trotting Stallion, American Eclipse,	323
The Three Friends,	330
Byron's Mazeppa,	335
Lady Suffolk,	341
Good for Heavy Drafts,	349
The Horse Tamed,	353
Bridle with a wooden Gag-bit for conquering vicious Horses,	358
Strap for the Right Fore-leg,	360
Strap for the Off Fore-leg,	362
Taming the Horse,	366
Teaching the Horse to lie down,	370
Struggles of the Vicious Horse against lying down,	373
Submission of the Horse,	377
Breaking the Horse to Harness,	380

THE ARAB AND HIS STEED.

History of the Horse

To man, whether as a civilized being or as a barbarian, no animal is more useful than the horse. The beauty, grace, and dignity of this noble creature, when in a properly developed state, are as marked as his utility. As an intelligent animal, he ranks next in the scale to the dog, that other companion and friend of man. Taking into consideration, then, his usefulness, his attractive appearance, and his intelligence, what is known of his history cannot prove unacceptable.

In order to ascertain the special land which can claim the proud honor of being the parent country, the birth-place of this noble animal, recourse must be had primarily to the pages of Scripture, as being the most ancient and best authenticated of all existing histories. By reference to those pages, we find that, although the ass was in early use among the children of Israel, the horse was unknown to them until after the commencement of their dwelling in Egypt; and strong evidence exists for the belief that he was not brought into subjection, even in that country, until after their arrival. Clear it is, at all events, that Arabia, which many have supposed to be the native home of the horse, did not possess him until within a comparatively recent period; while his introduction into Greece, and thence into those countries of Europe and Asia in which he is now found, either wild or domesticated, may be traced with much certainty to an Egyptian source.

Although in the history of Abraham frequent mention is made of the ass, of the camel, of flocks and herds, sheep and oxen, there is no allusion to the horse; nor, indeed, do we find any such until we reach the time of Joseph. In the reign of that Pharaoh in whose service Joseph was, wagons were sent by the king's command into Canaan, to bring thence into Egypt Jacob and his sons, their wives and their little ones, during the prevalence of the famine against which Joseph had provided. It is not recorded that those wagons were drawn by horses; but the inference that such was the fact is by no means irrational, when we remember that it was during the continuance of this famine that horses are first mentioned, having been taken by Joseph in exchange for bread from the Egyptian cultivators and cattle-breeders; that on the death of

Jacob, his funeral was attended by "both chariots and horsemen;" and lastly, that we know from the writings of Homer, and from the ancient sculptures of Persepolis and Nineveh, that the horse was used for purposes of draught for some time previous to his being ridden.

From this time, the horse appears to have been speedily adopted for use in battle. At the Exodus, some fifteen hundred years before the Christian era, the pursuing army contained "six hundred chosen chariots, and all the chariots of Egypt," together with all the horsemen. And when the Israelites returned into Canaan, we find that the horse had already been naturalized in that country, since the Canaanites "went out to fight against Israel with horses and chariots very many."

From these considerations, and from the fact that, so late as six hundred years after this date, Arabia had still no horses, it is by no means an improbable conclusion that the shepherd kings of Egypt, whose origin is unknown, introduced the horse into Lower Egypt; and that, after this period, that country became the principal herding district of this animal, whence he was gradually introduced into Arabia and the adjoining Asiatic countries. From the same stock is doubtless derived the entire race in all the southeastern parts of Europe. As Egypt is not, in any respect, a favorable country for horse-breeding, still less for his original existence in a state of nature, the source whence he was first introduced into that country is in some degree enveloped in uncertainty; though the better opinion, based upon much indirect testimony, is that he was an original native of the soil of Africa, which alone was the parent country of the Zebra and the Quagga—in some sort his kin.

It is questionable whether the horse is still to be found in a state of nature in Arabia; although it is asserted that they exist thinly scattered in the deserts, and that they are hunted by the Bedouins for their flesh, and also for the purpose of improving their inferior breeds by a different kind of blood. In central Africa, however, whence the horse is supposed to have been first introduced into Egypt, and thence into Arabia, Europe, and the East, wild horses still roam untamed far to the southward of the great desert of Sahara, where they were seen by Mungo Park in large droves.

At the period of the first Roman invasion, the horse was domesticated in Britain, and in such numbers, that a large portion of the forces which resisted the invaders were charioteers and cavalry.

In Europe, however, with but few exceptions, the horse, for purposes of warfare, was slowly, and not till the lapse of ages, brought into use: even the Spartans, the Athenians, and the Thebans, when at the height of their military renown, having but inferior and scanty cavalry services.

In the oldest sculptures probably in existence,—those removed by Layard from the ruins of Nineveh, and illustrative of almost every phase of regal and military life,—the horse is uniformly represented as a remarkably high-crested, large-headed, heavy-shouldered animal: rather long-bodied; powerfully limbed; his neck clothed with volumes of shaggy mane, often plaited into regular and fanciful braids; and his tail coarse and abundant, frequently ornamented similarly to his own mane and to the beard and hair of his driver—an animal, indeed, as unlike as possible to the low-statured, delicate-limbed, small-headed Arabs and barbs of modern days, with their basin-faces, large full eyes, and long, thin manes, from which the blood-horse of our times has derived his peculiar excellence. The same remarks may, in the main, be made as to the Greek and Roman horse, from the representations which have come down to us. The English blood-horse, being confessedly the most perfect animal of his race in the whole world, both for speed and endurance, and the American blood-

THE SHETLAND PONY.—AN ENGLISH SPORTING SCENE.

horse directly tracing without mixture to English, and through the English to Oriental parentage, some account of the former variety may be of interest to the reader.

It has already been remarked that large numbers of horses were found in Britain at the first Roman invasion. It is to be added, that Cæsar thought them so valuable that he carried many of them to Rome : and the British horses were, for a considerable period afterward, in great demand in various parts of the Roman Empire. After the evacuation of England by the Romans and its conquest by the Saxons, considerable attention was paid to the English breed of horses; and after the reign of Alfred, running horses were imported from Germany, this being the first intimation given us in history of running horses in England. English horses, after this, were so highly prized upon the Continent, that, in order to preserve the monopoly of the breed, in A.D. 930 a law was passed, prohibiting the exportation of the animal. In Athelstan's reign many Spanish horses were imported; and William the Conqueror introduced many fine animals from Normandy, Flanders, and Spain,—circumstances which show the strong desire, even at that early period, to improve the English breed. In the reign of Henry I. is the first account of the importation of the Arab horse into the country, at which time it is evident that the English had become sensible of the value and breed of their horse: and in the twelfth century a race-course had been established in London,—namely, Smithfield,—at once horse-market and race-course.

King John imported Flemish horses for the improvement of the breed for agricultural purposes; and in his reign is found the origin of the draught-horse now in general use in that country. Edward II. and Edward III. imported horses for the improvement of the stock, the latter introducing fifty Spanish horses. In the reign of Henry VII., the exportation of stallions was prohibited; but that of mares was allowed, when more than two years old, and under the value of six shillings and eight pence. In the reign of Henry VIII., many very arbitrary statutes were passed for the improvement of the horse; and it was during the same period that an annual race was run at Chester. In the reign of Elizabeth, the number and breed appear to have degenerated; for it is stated that she could collect but three thousand horse throughout her realm to resist the invasion of Don Philip.

With the accession of James I. to the throne, a great improvement was systematically wrought in the English breed; and from this period a constant and progressive attention was paid to the matter of breeding. This monarch purchased an Arabian horse at the then extraordinary price of five hundred pounds; but he proving deficient in speed, Arabians for a time fell into disrepute. Race meetings were then held at various places (Newmarket, among others) throughout the kingdom, the races being mostly matches against time, or trials of speed or bottom for absurdly long and cruel distances.

Although Cromwell, during his Protectorate, was obliged to forbid racing, yet he was an ardent lover of the horse, an earnest patron of all pertaining to horsemanship, and to his strenuous exertions the present superior condition of the English blood-horse is in no small degree owing.

Before proceeding to the history of the American horse—which is our main concern in the present branch of this work—a concise summary of the different varieties of this useful quadruped cannot fail to interest. We commence with the horse of Asia.

HORSES OF ASIA AND AFRICA.

THE ARABIAN.

In this country the horse, even in its wild state, (in which condition, as before remarked, it is rarely found,) is possessed of a beautiful symmetry of form, and a disposition of the greatest gentleness and generosity. His size is small, averaging in height generally between thirteen and fourteen hands, (the hand being reckoned at about four inches of our measure); color a dappled grey, though sometimes a dark brown; mane and tail short and black. The only mode of capturing him is by snares carefully concealed in the sand, as his exceeding swiftness prevents all possibility of taking him by the chase. The fondness of the Arab for his steed is well known, having long since passed into a proverb. The horse of the poorest wanderer of the desert shares with his master and his family every attention and caress which the strongest attachment can prompt. Mares are always preferred by the Arab to horses, as they endure fatigue and the hardships incident to a desert life much more patiently, and they can be kept together in greater numbers without the risk of quarrels and mutual injuries. Great attention is paid to the coat of the animal. He is carefully washed each morning and evening, or after a long ride; is fed only during the night, receiving throughout the day nothing but one or two drinks of water.

The head of the pure Arab is light, well made, wide between the nostrils, forehead broad, muzzle short and fine, nostrils expanded and transparent, eyes prominent and sparkling, ears small; neck somewhat short; shoulders high and well thrown back; withers high and arched; legs fine, flat, and small-boned, and the body somewhat light.

THE PERSIAN.

This horse is slightly taller than the Arab: is full of bone, and very fast. The Persian feeds his horse as does the Arab, the food given being coarse and scant. Hay is utterly unknown for the purpose, barley and chopped straw being generally substituted. Although this variety is in most respects less esteemed than the Arab, it is in some points its superior.

THE TARTARIAN.

Like the Persian, this variety is swift; but the horses are heavy-headed, low-shouldered, and altogether very awkwardly put together. The Tartars eat the flesh of their horses and use the milk of their mares, from which they also make excellent cheese.

THE TURKOMAN.

This is a variety of the Tartar, but superior to it; bringing, even in Persia, frequently from five hundred to a thousand dollars. Its average height is some fifteen hands, and in general appearance it bears a strong resemblance to a well-bred English carriage-horse. Though possessed of considerable speed, it is not enduring. This variety is often foisted upon the ignorant as the pure Arabian.

THE TURKISH HORSE.

This horse is a cross between the Persian and the Arabian, and is of slender build, carrying his head high, lively and fiery, and possessing a gentle and affectionate disposition. The tail of the horse is regarded in Turkey and Persia as a badge of dignity, princes measuring their rank by the number of tails they carry; those of the highest rank being allowed three.

HORSES OF HINDOSTAN.

In India, the horse, owing to the peculiar climate of the country, is invariably found to degenerate, unless great attention be paid to breeding. The principal breeds are the Tazee, the Takan, the Folaree, the Cutch, and the Dattywarr.

Passing from the Asiatic horses to the African, it is to be remarked that Egypt has long since lost its character as a breeding country, its horses being justly deemed much inferior to those of Persia, Barbary, or Arabia.

THE BARB, AND OTHERS.

This variety—the principal of the African race—is taller than the Arabian, and is remarkable for the height and fullness of its shoulders, drooping of the haunches, and roundness of the barrel.

The Bornou race, in the central parts of Africa, is described as possessing the qualities of the Arabian with the beauty of the Barb; as being fine in shoulder and of general elegance of form. The Nubian horses are stated by travelers to be even superior to the Arabian. Dongola has a noticeable breed, of large size, their chief peculiarities being extreme shortness of body, length of neck, height of crest, and a beautiful forehand.

EUROPEAN HORSES.

THE RACER.

As the varieties of the horse in Great Britain are the most noticeable of any in Europe, we append a brief description of the principal breeds at present in use.

The Racer, which excels, in beauty, speed, and endurance, that of all other nations, was gradually formed by the introduction of the best blood of Spain, Barbary, Turkey, and Arabia, and bears a strong family likeness to each. The characteristics of this breed are a high and lofty head, bright and fearless eye, small ear, expanded nostril; arched neck, curved on the upper surface, with no curve underneath; the neck gracefully set on; the shoulder lengthened, oblique, and lying well back; the quarters ample and muscular; the fore-legs straight and fine, but with sufficient bone; the hinder legs well bent, and the pasterns long and springy.

THE HUNTER.

The best horses of this breed stand fifteen or sixteen hands high: head small; neck thin, especially beneath the crest, firm and arched; and jaws wide; lofty forehead; shoulders as extensive and oblique as that of the racer, and somewhat thicker; broad chest; muscular arm; leg shorter than that of the racer; body also more short and compact; loins broad; quarters long; thighs muscular; hocks well bent, and under the horse.

THE HACKNEY.

This horse is still more compact than the hunter, with more

substance in proportion to his height; forehead light and high; head small, and placed taperingly upon the neck; shoulders deep and spacious, lying well back; back straight, loins strong; fillets wide, and withers well raised. Too high breeding is considered objectionable in this species, as being ill adapted for ordinary riding upon the road.

THE CART HORSE.

The principal varieties of this class, are the Cleveland, the Clydesdale, the Northamptonshire, the Suffolk Punch, and the heavy black or dray horse. The Clydesdale breed obtains its name from being bred chiefly in the valley of the Clyde. They are strong and hardy, have a small head, are longer necked than the Suffolk, with deeper legs and lighter bodies. The Suffolk Punch

THE STALLION.

originated by crossing the Suffolk cart mare with the Norman stallion. Its color is yellowish or sorrel; large head, wide between the ears, muzzle rather coarse, back long and straight, sides flat, fore-end low, shoulders thrown much forward, high at the hips, round legs, short pasterns, deep-bellied, and full barrel. The modern-bred cart horse of England, originated from a cross with the Yorkshire half-bred stallion, and is of much lighter form, and stands much higher. This horse is hardy and useful, kindly, and a good feeder. The heavy black horse is chiefly bred in Lincolnshire and the Midland counties.

GERMAN, FRENCH, AND SPANISH HORSES.

The horses of Germany, with the exception of the Hungarian, are generally large, heavy, and slow. The Prussian, German, and the greater part of the French cavalry, are procured from Holstein. They are of a dark glossy bay color, with small heads, large nostrils, and full dark eyes, being beautiful, active, and strong.

The horses of Sweden and Finland are small, but beautiful, and remarkable for their speed and spirit; those of Finland being not more than twelve hands high, yet trotting along with ease at the rate of twelve miles an hour.

The Iceland horse is either of Norwegian or Scottish descent. They are very small, strong, and swift. Thousands of them live upon the mountains of that barren country, never entering a stable, but taught by instinct or habit to scrape away the snow, or break the ice, in quest of their meagre food.

The Flemish and Dutch horses are large, and strongly and beautifully formed. The best blood of draught horses is owing, in a great degree, to crosses with these.

The best French horses are bred in Limousin and Normandy; the provinces of Auvergne and Poitou producing ponys and galloways, which are excellent saddle-horses and hunters.

The Spanish horse of other days, as the Andalusian charger

and the Spanish jennets, exists but in history or romance. The modern Spanish horse resembles the Yorkshire half-bred, with flatter legs and better feet, but a far inferior figure.

The Italian horses, particularly the Neapolitan, were once in high repute; but, owing mainly to intermixtures of European, rather than Eastern blood, they have sadly degenerated.

THE AMERICAN HORSE.

At a very remote period in the history of America, the horse began to be imported from Europe by the earliest settlers; it being conceded that, although the horse had, at some former time, existed on this continent, as is proved by his fossil remains, which have been found in abundance in various parts of the country, he had become extinct previous to its colonization by the white nations.

AMERICAN FARM SCENE.

It is generally believed that the horses which are found in a wild state on the *pampas* or plains of South America, and the prairies of North America, as far east as to the Mississippi River, are the descendants of the parents set loose by the Spaniards at the abandonment of Buenos Ayres. This opinion, however, is combated by some, on the ground that this date is too recent to account for the vast numerical increase, and the great hordes of these animals now existing in a state of nature; and they are inclined to ascribe their origin to animals escaped, or voluntarily set at liberty, in the earlier expeditions and wars of the Spanish invaders, the cavalry of that nation consisting entirely of perfect horses or mares.

An opportunity for such an origin must undoubtedly have been furnished in the bloody wars of Mexico and Peru; since upon the issue of many battles, which were disastrous to the Spaniards, the war-horses, their riders being slain, could have recovered their freedom and propagated their species rapidly in the wide, luxuriant, and well-watered plains, where the abundance of food, the genial climate, and the absence of beasts of prey capable of successfully contending with so powerful an animal as the horse, would favor their rapid increase.

We know, moreover, that De Soto had a large force of cavalry in that expedition in which he discovered the Mississippi, and found a grave in its bosom; and when his warriors returned home in barques which they built on the banks of the "Father of waters," there can be little doubt that their chargers must have been abandoned, since their slender vessels, built by inexperienced hands for the sole purpose of saving their own lives, must have been incapable of containing their steeds.

The first horses imported to America for the purpose of creating a stock, were brought by Columbus, in 1493, in his second voyage to the islands. The first landed in the United States, were introduced into Florida in 1527, by Cabeca de Vaca, forty-two in number; but these all perished or were killed. The next importation was that of De Soto, before mentioned, to which is doubtless to be attributed the origin of the wild horses of Texas and the prairies, a race strongly marked to this day by the characteristics of Spanish blood.

In 1604, L'Escarbot, a French lawyer, brought horses and other domestic animals into Acadia; and in 1608, the French, then engaged in colonizing Canada, introduced horses into that country, where the present race, though somewhat degenerated in size, owing probably to the inclemency of the climate, still shows the blood, sufficiently distinct, of the Norman and Breton breeds.

In 1609, the English ships landing at Jamestown, in Virginia, brought, besides swine, sheep, and cattle, six mares and a horse; and in 1657, the importance of increasing the stock of this valuable animal was so fully recognized, that an act was passed, prohibiting its exportation from the province.

In 1629, horses and mares were brought into the plantations of Massachusetts Bay, by one Francis Higginson, formerly of Leicestershire, England, from which county many of the animals were imported. New York first received its horses in 1625, imported from Holland by the Dutch West India Company, probably of the Flanders breed, though few traces of that breed yet exist, unless they are to be found in the Conestoga horse of Pennsylvania, which shows some affinity to it, either directly or through the English dray-horse, which latter is believed to be originally of Flemish origin.

In 1750, the French of Illinois procured considerable numbers of French horses; and since that time, as the science of agriculture has improved and advanced, pure animals of many distinct breeds have been constantly imported into this country, which have created in different sections and districts distinct families, easily recognized,—as the horses of Massachusetts and Vermont, admirable for their qualities as draft-horses, powerful, active, and capable of quick as well as heavy work; the Conestogas, excellent for ponderous, slow efforts, in teaming and the like; and the active, wiry horses of the West, well adapted for riding, and being in most general use for American cavalry purposes.

It is evident, then, that the original stock of the unimproved American horse is the result of a mixture of breeds; the French, the Spanish, the Flemish, and the English horses having all sent their representatives to some one portion at least, of the United States and British Provinces, and probably still prevailing to a considerable degree in some locations, though nowhere wholly unmixed—while, in others, they have become so thoroughly mixed and amalgamated, that their identity can no longer be discovered.

In New York, for example, the early importations of thorough blood, and the constant support of horse-racing, appear to have so changed the original Dutch or Flemish stock, that the characteristic of her horses is that of the English race, with a decided admixture of good blood. In Massachusetts, Vermont, and the Eastern States generally, the Cleveland bay, and a cross between that and the English dray-horse blood, with some small admixture of thorough blood, predominate. In Pennsylvania, the most distinct breed appears to be of Flemish and English dray-horse origin. In Maryland, Virginia, and South Carolina, English thorough blood prevails to a great extent; so much so as to render the inferior class of working horses undersized. In Louisiana, and many of the

Western States, French and Spanish blood partly prevail, though with a mixture of English blood. It may, in short, be generally assumed that, with the exception of the thorough-breds, there is scarcely any breed in any part of America wholly pure and unmixed; and that there are very few animals anywhere, which have not some mixture, greater or less, of the hot blood of the East, transmitted through the English race-horse.

Indeed, with the exception of the Conestoga horse, there is, in the United States, no purely-bred draft or cart-horse, nor any breed which is kept entirely for labor in the field or on the road, without a view to being used at times for quicker work, and for purposes of pleasure or travel. Every horse, for the most part, bred in America, is intended to be, in some sense, used upon the road; and it is but asserting a well-known fact, when we say, that for docility, temper, soundness of constitution, endurance of fatigue, hardiness, sure-footedness, and speed, the American roadster is not to be excelled, if equaled, by any horse in the entire world not purely thorough-bred.

Of roadsters, two or three families have obtained, in different localities, decided reputations for different peculiar qualities: such as the Narragansett pacers, the families known as the Morgan and Black Hawk, the Canadian, and generally what may be called trotters. No one of these, however, with the single exception of the Narragansetts, appears to have any real claim to be deemed a distinctive family, or to be regarded as capable of transmitting its qualities in line of hereditary descent, by breeding within itself, without further crosses with higher and hotter blood.

Of the Narragansetts, but little can be said with certainty;

for there is reason to believe that, as a distinct variety, with natural powers of pacing, they are extinct; and their origin is, in some degree, uncertain. The other families clearly owe their merits to a remote infusion of thorough-blood, perhaps amounting to one-fourth, or one-third part, some three or four generations back.

The original Canadians were, doubtless, of pure Norman and Breton descent; but, since the Canadas have been under British rule, they also

THE CANADIAN.

have been largely mixed with, and much improved by, the introduction of a pure blood; so that the animals which in late years pass here under the name of Canadians, such as Moscow, Lady Moscow, and many others of that name, are Canadians only in name, differing from other American roadsters simply in the fact that they have, for the most part, only two crosses of the Norman and pure English blood, while the ordinary road-horse of the United States is perhaps a combination of several distinct English families, with French, Spanish, and Flemish crosses, besides an infusion of thorough-blood.

Of trotters, there is certainly no distinctive breed or family, or mode of breeding. The power, the style, the action, the mode of going, are the points regarded; and it is most probable, that the speed and the endurance, both of weight and distance, depend, more or less, on the greater or inferior degree of blood in the animal.

Indeed, the wonderful superiority of the American roadster is attributable to the great popularity of trotting in this country, to the great excellence of the trotting-trainers, drivers, and riders, arising from that popularity, and to the employment of all the very best half and three-quarter-part bred horses in the land for trotting purposes, none being turned from that use for the hunting-field or park-riding.

The general American horse, as compared with the English horse, is inferior in height of the forehand, in the loftiness and thinness of the withers, and in the setting-on and carriage of the neck and crest; while he is superior in the general development of his quarters, in the let-down of his hams, and in his height behind; and further remarkable for his formation, approaching what is often seen in the Irish horse, and known as the goose-rump. Even the American racer stands very much higher behind and lower before than his English fellow.

Another point in which the American horse of all conditions differs extremely and most advantageously from the European animal, is his greater sure-footedness and freedom from the dangerous vice of stumbling. Any one can satisfactorily convince himself of this, by comparing the knees of hack-horses let for hire, either in the cities or rural villages of the United States, with those of similar English localities. In this country, a broken knee is one of the very rarest blemishes encountered in a horse; while of horses let for hire in England, with the exception of those let by a few crack livery-keepers in London, in the Universities, and in one or two other of the most important towns in hunting neighborhoods, a majority are decidedly broken-kneed.

The exemption of the horse, on this side of the Atlantic, from this fault, is ascribable: first, to the fact, that both the pasture-lands and the roads here are far rougher, more broken in surface, and more interrupted by stones, stumps, and other obstacles, than in the longer cultivated and more finished countries of Europe, which teaches young horses to bend their knees, and throw their legs more freely while playing with the dams in the field; and also to lift and set down their feet with much greater caution even on our great thoroughfares; secondly, to the higher blood and breed of riding-horses in England, which are often cantering thorough breds, liable to be unsafe travelers on the road; and lastly, to the well-known circumstance, that most of the hired horses are roadsters—these are worn-out or broken-down animals of a higher caste, which are deemed, by reason of their disqualification for a higher position, fit for a secondary one, although suited to none, and dangerous in any.

To this admirable quality of the American horse, must be added his extreme good temper and docility, in which he undeniably excels any other horse in the world. From the first childhood of the animal until he is fully put to work, he requires and receives little or no breaking, unless he show qualities which promise such speed or endurance as to render it advisable to train him as a trotter. Even when this is done, it is for the purpose of developing his powers, getting him to exert himself to the utmost, and teaching him how to move to the best advantage; and not to render him submissive, easy of management, or gentle to be handled. There is scarcely ever any difficulty in saddling, in harnessing, in backing, or in inducing him to go. He may be awkward at first, uncouth, shy, and timid; but he is never, one may almost say, violent, spasmodic in his actions, and fierce.

It is true that horses are treated, for the most part, with superior judgment and greater humanity in the United States; that the whip is little used, and the spur almost unknown; still the whole of this remarkable difference in temper, on the part of the American horse, cannot be attributed to the difference of treatment.

As he begins, moreover, he continues to the end. One rarely encounters a kicker, a runaway, an inveterate shyer or balker, and hardly ever a furious animal, not to be approached, save at the risk of limb or life, in an American horse of any class or condition.

Probably this fact may, in some respects, be attributed to the less high strain of blood in the American roadster, and still more to the hardier and less stimulating mode of treatment to which he is subjected. The heating treatment to which the English horse is subjected, unquestionably deprives him, in some degree, of the power of enduring long-protracted exertion, privation, hardship, and the inclemency of the weather; and the pampering, high feeding, excessive grooming, and general maintenance of horses in an unnatural and excited state of spirits has, assuredly, an injurious effect upon the general temper of the animal; though not, perhaps, so greatly as to account for all the difference to which allusion has just been made.

Having premised thus much, in general terms, of the history and peculiarities of the general American horse, we will next take up the leading varieties which obtain in this country, commencing with

THE AMERICAN BLOOD-HORSE.

Unlike the human race of the United States, unlike the ordinary working horse, unlike the cattle and most of the domestic animals of North America—which cannot be traced or said to belong to any single distinct breed or family, having originated from the combination and amalgamation of many bloods and stocks, derived from many different countries—the blood-horse, or racer, of America stands alone, unquestionably of pure English thorough-blood.

What that English thorough-blood is, it is only necessary here to say that, although it is not possible, in every instance, to trace the great progenitors of the English and American turf, directly on both sides, to Desert blood; and although it can scarcely be doubted that, in the very commencement of turf-breeding, there must have been some mixture of the best old English blood, probably, in great part, Spanish by descent, with the true Arab or Barb race; yet the impure admixture is so exceedingly remote, not within fourteen or fifteen generations—since which the smallest taint has been carefully excluded—that the present race-horse of England or North

America, cannot possess above one sixteen-thousandth part of any other blood than that of the Desert.

Nor can it be doubted, that the modern thorough-bred is far superior to the present horse of the East in his qualities and powers, as he is in size, bone, strength, and ability to carry weight. It is to this very superiority of our thorough-bred, which has been proved wherever it has encountered the Oriental horse, that it must be ascribed, that no late cross of Arab blood has, in the slightest degree, improved the European or American racer.

It seems now to be a conceded point, that to improve any blood, the sire must be the superior animal; and, since by care, cultivation, superior food, and better management, our descendant of Desert blood has been developed into an animal superior to his progenitors, mares of the improved race can gain nothing by being crossed with the original stock; although it is yet to be seen, whether something might not be effected by the importation of Oriental mares, and breeding them judiciously to modern thorough-bred stallions.

It has been already stated, that the first systematic attempts at improving the blood of the English horse began in the reign of King James I., was continued in that of Charles I. and during the Commonwealth, and advanced with renewed spirit on the restoration of the Stuarts. In the reign of Queen Anne, the last of that house who occupied the English throne, the English thorough-bred horse may be regarded as fairly established; the Darley Arabian, sire of Flying Childers, Curwen's Barb, and Lord Carlisle's Turk, sire of the Bald Galloway, being imported in her reign. Sixteen years after her death, and three years before the settlement of Georgia, the youngest of the original American colonies, twenty-one foreign, and fifty native stallions, some of them the most celebrated horses which the world has ever seen, were in service as stock-getters in the United Kingdom; and from some of these are descended all our racers of the present day.

It was precisely during this period that the American colonies were planted; and, as might be anticipated, English horses of pure blood were introduced at a very early date. Indeed, in those sections where the settlement was mainly effected by men attached to the Cavalier party, race-horses were kept and trained, race-courses were established, and a well-authenticated stock of thorough-bred animals, tracing to the most celebrated English sires, many of which were imported in the early part of the eighteenth century, was in existence for some time before the outbreak of the old French war.

In the Eastern States, whose settlers were mainly attached to the Puritan party, and therefore opposed in an especial manner to horse-racing, very few horses of thorough blood were imported.

In Virginia and Maryland, as the head-quarters of the Cavaliers, it is probable that racing commenced simultaneously, or nearly so; it being an attribute of the principal towns of Maryland some years prior to Braddock's defeat in 1753. In the latter State, indeed, it appears for some time to have been considered a part of the duty of the Governor to keep a racing stud; since no less than five successive governors were all determined turfmen and supporters of the American racing interest.

As our Revolutionary War interrupted the peaceful progress of the country and the avocations of our country gentlemen at so early a period in the history of the American Turf, the difficulty of ascertaining how far records or registries have been preserved, or were kept from the first, has been materially enhanced. Yet, on the whole, it may be regarded as remarkable rather that so many pedigrees can be unequivocally followed out, than that a few should be obscure and untraceable farther than to an imported mare. Indeed, it must be granted as a fact which cannot be questioned or doubted, fully established both by their own performances and by the unfailing transmission of their hereditary qualities, that our American horses are as certainly thorough-bred as are any of those English champions, whose blood no one ever dreams of disputing, which go back, like that of Eclipse himself, or many others of equal renown, to an unknown dam or sire.

From Virginia and Maryland, the racing spirit extended rapidly into the Carolinas, where it has never to this day flagged. The oldest race-courses in this country, which are yet kept up for purposes of sport, are the Newmarket course, near Petersburg, Virginia; and the Washington course, near Charleston, South Carolina. At Alexandria, D. C., there was a race-course early in the last century, and the courses in the neighborhood of Richmond have been in existence above seventy years.

It was not until about the commencement of the present century, that what may be called race-courses proper were established in New York; the first club for the promotion of the breed of horses by means of racing dating from 1804; although long previously the improvement of the breed of horses had created much interest in that State, celebrated stock-getters having been imported as early as 1764 and 1765.

Into Pennsylvania, a State which has never particularly distinguished itself in the racing turf, were brought at an early date two horses, Gray Northumberland (also called Irish Gray), and Old England; to these must be given the credit of running one of the oldest great American time-races on record, as long ago as 1767, against two other horses, Selim and Granby.

Although the use of the horse for merely racing purposes does not at present obtain to as great an extent with us as in England—a circumstance which can be readily accounted for from the prejudice which many entertain

BLACK HAWK—AN AMERICAN RACER.

against such a use, owing to the objectionable accompaniments which are too often found in connection with it—still it should not be forgotten, that the advantage to be derived from the thorough-bred horse depends upon far more than his applicability to the turf and his fitness for racing purposes. Were it otherwise, it would scarcely be worth while to devote the space to the consideration of this topic which has, by common consent, been deemed indispensable.

The truth is, that the race-course was not, in the beginning, so much as thought of as a scene for the display of the high qualities of this animal; much less was racing considered by our ancestors as an end for which they imported the Eastern horse into Europe. It was for the improvement of the native stock of horses in the various European Kingdoms, by giving to them speed and endurance,—in which respects no other breed can compare with them,—that the Asiatic and North-African horse was so eagerly sought by the monarchs, especially of England, during the seventeenth, and the early part of the eighteenth century.

The race-course was at first employed solely as a method of testing the prevalence or superiority, in certain animals or breeds of animals, of these qualities of speed and endurance, which can by no other known method be so completely, so accurately, and so fairly tested. Soon after the introduction of the thorough-bred horse, this process of testing his qualities grew into a favorite sport with all classes of persons in England. After the multiplication of race-courses throughout the kingdom and the establishment of racing as a national institution, the objects of the possessors and breeders of race-horses underwent a change: what had been a means originally, becoming eventually, more or less, the end. Horses, in a high form and of the most favorite and purest strains of blood, were eagerly sought and commanded large prices, for the purposes of sport and honorable competition, as was the case in the Olympic Games of ancient Greece.

At a yet later date, a second change of object has taken

place; and, with but few exceptions, the thorough-bred horse is now kept, both in England and this country, for the paramount purpose of money-making, either by the actual winning of his prizes, or by his service in the stud, after his racing career is ended.

Still, although the animals employed may be generally kept merely for the gratification of cupidity and the excitement of the contest, and though racing and race-courses may be subject to abuses by far too many, yet such means are, even now, as they were intended to be from the first, the best and only mode of really improving the general stock of any country. As the points of the thorough-bred horse are precisely those which constitute the perfection of a blood-horse in a high form as a stallion for improving the breed of animals, and for getting the best horses from any possible class of mares, for all possible uses, unless for the very slowest and most ponderous draught, the description of those points which are most generally accepted as accurate is subjoined.

Purity of blood is an indispensable requisite for the thorough-bred horse. By the term "blood," it is not intended to be understood that there is any real difference between the blood of the thorough-bred horse and that of the half-bred animal, as no one could discriminate between the two by any known process. The term is here used in the same sense as "breed," and by purity of blood is meant purity in the breeding of the individual animal under consideration; that is to say, that the horse which is entirely bred from any one source is pure, or free from any mixture with others, and may be a pure Suffolk Punch, or a pure Clydesdale, or a pure thorough-bred horse. All these terms are, however, comparative, since there is no such animal as a perfectly purely-bred horse of any breed, whether cart-horse, hack, or race-horse; all have been produced from an admixture with other breeds, and though now kept as pure as possible, yet they were originally compounded of varying elements. As, however, the thorough-bred horse as he is called, has long been bred for racing purposes, and selections have been made with that view alone, it is reasonable to suppose that this breed is the best for that purpose, and that a stain of any other is a deviation from the classical stream into one more muddy, and therefore impure. Indeed, in actual practice this is found to be the case; for in every instance it has resulted that the horse bred with the slightest deviation from the sources indicated by the stud-book, is unable to compete in lasting power with those which are entirely of that breed. Hence it is established as a rule, that for racing purposes every horse must be thorough-bred; that is, bred of a sire and dam, whose names are found in the stud-book.

The external form of the blood-horse is of great importance; it being true, other things being equal, that the horse will be the best runner which is formed in the mould most like that of the greatest number of good race-horses. Still, it is admitted on the turf, that high breeding is of more consequence than external shape, and that, of two horses, one perfect in shape but of an inferior strain of blood, and the other of the most winning blood, but in shape not so well formed, the latter will be the most likely to give satisfaction on the race-course. Hence originates the proverb, "an ounce of blood is worth a pound of bone." Yet, in spite of all this recognized superiority of blood, it is indisputable that for the highest degree of success there must be not only high purity of blood, and that of the most winning strains, but there must also be a frame of the most useful character, if not always of the most elegant form. Many of our very best horses have been plain and even coarse-looking; but, in spite of their plainness, all their points were good and useful, and the deficiency was in mere elegance, not in real utility.

The height of the race-horse varies from fifteen hands to sixteen and a half hands, or even seventeen hands; but the general height of our best horses, is about fifteen hands and three inches. Few first-class performers have exceeded the height of sixteen hands and one inch. The average, above given, may be fairly laid down as the best height for the race-horse; though it cannot be denied, that for some small and confined courses, a smaller horse, of little more than fifteen hands high, has a better chance, as being more capable of turning round the constantly recurring angles or bends.

The head and neck should be characterized by lightness, which is essential for this department. Whatever is unnecessary is so much dead weight; and whatever is found in the head and neck, which is not necessary for the peculiar purposes of the race-horse, is

AMERICAN PLANTATION SCENE.

so much weight thrown away, which must still be carried by the horse. The head, we may say in detail, should be lean about the jaw, yet with a full development of forehead, which should be convex and wide, so as to contain within the skull a good volume of brain. If this fullness exist, all the rest of the head may be as fine as possible; the jaws being reduced to a fine muzzle, with a slight hollowing out in front, but with a width between the two sides of the lower jaw where it joins the neck, so as to allow plenty of room for the top of the windpipe when the neck is bent. The ears should be pricked and fine, but not too short; eyes full and spirited; nostrils large, and capable of being well dilated when at full speed, which is easily tested by the gallop, after which they ought to stand out firmly, and so as to show the internal lining fully. The neck should be muscular, and yet light; the windpipe loose and separate from the neck,—that is, not too tightly bound down by the membrane of the neck. The crest should be thin and wiry, not thick and loaded, as is often seen in coarse stallions, or even in some mares.

Between the two extremes of the ewe-neck and its opposite, there are many degrees, but for racing purposes the former is preferable of the two, to the latter; for few horses can go well with their necks bent so as to draw the chin to the bosom; yet here, as in other cases, the happy medium between the two extremes is the most desirable.

The body, or middle-piece, should be moderately long, and not too much confined between the last rib and the hip bone. So long as the last or back-ribs are deep, it is not of so much importance that they should be closely connected to the hip-bone, for such a shape shortens the stride; and though it enables the horse to carry a great weight, yet it prevents him from attaining a high rate of speed. The back itself should be muscular, and the hips so wide as to allow of a good development of the muscular department. The withers may rise gently, but not too high, with that thin, razor-like elevation which many people call a good shoulder, but which really has nothing to do with that part, and is only an annoyance to the saddler, in preventing its being pinched by the saddle. The chest itself should be well developed, but not too wide and deep; no horse can go a good distance without a fair "bellows-room;" but, supposing the beast to be sound and of good quality, the amount of lungs will suffice which may be contained in a medium-sized chest, and all above that is wasted, and is extra weight. Many of our best-winded horses have had medium-sized chests, and some of the very worst have been furnished with room enough for a blacksmith's bellows to play in. If the heart only does its duty well, the lungs can always furnish sufficient air; and we know that when frequently renewed, and with sufficient power, the blood is aerated as fast as it is propelled, and the chief difficulty lies in this power of propulsion, which resides in the heart alone. If the chest be too wide, it materially affects the action of the fore-legs, and, therefore, in every point of view, theoretically and practically, there is a happy medium between the too great contraction in this department, and the heavy, wide, lumbering chests, sometimes seen even in the thorough-bred race-horse, especially when reared upon rich succulent herbage, more fitted for the bullock than for the Eastern horse. In the formation of the hips, the essential point is length and breadth of bone for muscular attachment, and it matters little whether the croup droops a little, or is pretty straight and level, so that there is a good length from the hips to the haunch-bone the line between which two points may be either nearly horizontal, or forming a considerable angle with the ground; but still in both cases it should be a long line, and the longer it is the more muscular substance is attached to it, and the greater leverage will the muscles have.

The fore-quarter, consisting of the shoulder, upper and lower arm and leg and foot, should be well set on to the chest; and the shoulder-blade should lie obliquely on the side of that part, with a full development of muscle to move it, and thrust it well forward in the gallop. Obliquity is of the greatest importance, acting as a spring in taking off the shock of the gallop or leap, and also giving a longer attachment to the muscles, and in addition enabling them to act with more leverage upon the arm and leg. As the shoulder-blade does not reach the top of the withers, and as the bones forming that part have nothing to do with the shoulder itself, many high-withered horses have bad and weak shoulders, and some very upright ones; whilst, on the other hand, many low-withered horses have very oblique and powerful shoulders, and such as to give great facility and pliability to the fore extremity. The shoulder should be very muscular, without being over-done or loaded, and so formed as to play freely in the action of the horse. The point of the shoulder which is the joint corresponding to the human shoulder, should be free from raggedness, but not too flat; a certain degree of development of the bony part is desirable, but more than this leads to defects, and impedes the action of this important part. The upper arm, between this joint and the elbow, should be long, and well clothed with muscles; the elbow set on quite straight, and not tied in to the chest; the lower arm muscular and long; knees broad and strong, with the bony projection behind well developed; legs flat, and showing a suspensory ligament large and free; pasterns long enough, without being weak; and the feet sound, and neither too large nor too small, and unattended with any degree of contraction, which is the bane of the thorough-bred horse.

The hind-quarter is the chief agent in propulsion, and is therefore of the utmost consequence in attaining a high speed. It is often asserted that the oblique shoulder is the grand requisite in this object, and that it is the part upon which speed mainly depends, and in which it may be said to reside. This is, to some extent, true, because there can be no doubt that with a loaded shoulder high speed is impracticable; for, however powerfully the body may be propelled, yet when the fore-quarter touches the ground it does not bound off again as smartly as it ought to do, and the pace is consequently slow. The elastic shoulder, on the contrary, receives the resistance of the earth, but reacts upon it, and loses very little of the power given, by the strike of the hind-quarter, which, nevertheless, must be strong and quick, or else there is nothing for the shoulder to receive and transmit. For the full action of the hind-quarters, two things are necessary, viz: first, length and volume of muscle; and, secondly, length of leverage, upon which that muscle may act. Hence, all the bones comprising the hind-quarter should be long, but the comparative length must vary a good deal, in order that the parts upon which the muscles lie may be long, rather than those connected with the tendons, which are mere ropes, and have no propelling power residing in them, but only transmit that which they derive from the muscles themselves. Thus, the hips should be long and wide, and the two upper divisions of the limb—viz., the stifle and lower thigh—should be long, strong, and fully developed. By this formation, the stifle-joint is brought well forward, and there is a considerable angle between these two divisions. The hock should be long and strong, free from gum or spavin, and the point long, and so set on as to be free from weakness at the situation of curb. In examining the hind-quarter, to judge of its muscular development, the horse should not be looked at sideways, but his tail should be raised, and it should be ascertained that the muscles of the two limbs meet together below the *anus*, which should in fact be well supported by them, and not left loose, and, as it were, in a deep and flaccid hollow. The outline of the outer part of the thigh should be full, and in ordinary horses the muscles should swell out beyond the level of the point of the hip. This fullness, however, is not often seen to such an extent in the thorough-bred horse, until he has arrived at mature age, and is taken out of training. The bones below the hock should be flat and free from adhesions: the ligaments and tendons fully developed, and standing out free from the bones; and the joints well formed and wide, yet without any diseased enlargement; the pasterns should be moderately long, and oblique; the bones of good size; and, lastly, the feet should correspond to those already alluded to in the anterior extremity.

These points, taken as a whole, should be in proportion to one another—that is to say, the formation of the horse should be "true." He should not have long, well-developed hindquarters, with an upright, weak, or confined fore-quarter. Nor will the reverse of this answer the purpose; for, however wellformed the shoulder may be, the horse will not go well unless he has a similar formation in the propeller. It is of great importance, therefore, that the thorough-bred horse should have all his various points in true relative development, and, that there should not be the hind-quarter of a long, racing-like horse, with the thick, confined shoulder which would suit a stride less reaching in its nature.

The color of the thorough-bred horse is now generally bay, brown, or chestnut, one or the other of which will occur in ninety-nine cases out of a hundred; gray not being common, though it sometimes appears. Black, also, occasionally makes its appearance, but not more frequently than gray. Roans, duns, sorrels, etc., are now quite exploded, and the above five colors may be said to complete the list of colors seen in the race-horse. Sometimes these colors are mixed with a good deal of white, in the shape of blazes on the face, or white legs and feet; or even both may occur, and the horse may have little more than his body of a brown, bay, or chestnut. Most people, however, prefer the self-color, with as little white as possible; and nothing but the great success of a horse's stock would induce breeders to resort to him, if they were largely endowed with white. Gray hairs mixed in the coat, are rather approved than otherwise; but they do not amount to a roan, in which the gray hairs equal, or even more than that, the other colors mixed with them.

The texture of the coat and skin is a great proof of high-breeding, and, in the absence of the pedigree, would be highly regarded; but when that is satisfactory, it is of no use descending to the examination of an inferior proof; and, therefore, except as a sign of health, the skin is seldom considered. In all thorough-bred horses, however, it is thinner, and the hair more silky than in common breeds; and the veins are more apparent under the skin, partly from its thinness, but also from their extra size and number of branches. This network of veins is of importance in allowing the circulation to be carried on during high exertions, when, if the blood could not accumulate in them, it would often choke the deep vessels of the heart and lungs; but, by collecting on the surface, great relief is afforded, and the horse is able to maintain such a high and long-continued speed, as would be impracticable without their help. Hence these points are not useful as a mere mark of breed, but as essential to the very purpose for which that breed was established.

The mane and tail should be silky, and not curly, though a slight wave is often seen. A decided curl is almost universally a mark of degradation, and shows a stain in the pedigree as clearly as any sign can do. Here, however, as in other cases, the clear tracing of that all-powerful proof of breeding, will upset all reasoning founded upon inferior data. The setting on of the tail is often regarded as of great importance, but it is chiefly with reference to appearances; for the horse is not dependent for action or power upon this appendage. Nor is strength of dock of any value as a sign, and many very stout horses have been known with flaccid and loosely pendant tails.

It is well known that certain horses can run half a mile at high speed, but no more; others, a mile; others, again, a mile and a half, or two miles; whilst another class, now less common than formerly, require a distance of three or four miles to develop their powers, as compared with ordinary horses. These peculiarities are generally hereditary, though not always so; but still, when the blood is known, it may generally be surmised, that the individual will, or will not, stay a distance. When the cross in question is stout on one side, and flashy on the other, it is not easy to guess to which the young scion may lean; but in those cases where a horse is bred from sire or dam, both of stout blood, or the reverse, the experienced hand may, in almost all cases, decide beforehand upon the qualities of the son or daughter, as far as staying qualities are concerned. Again, there are some horses of strong, compact frames, with short backs and strong quarters, who may be expected to climb a hill without difficulty, especially if of stout blood; and, again, there are others of lathy frames, with long but weak points, and a great deal of daylight under them, who may win over the flat for a mile, or a mile and a quarter, but can never climb a hill, or get beyond the above distance over a flat. All these points should be carefully studied by the breeder in getting together his breeding stock, and by the owner in deciding how he will enter his young produce in the race.

In passing from the consideration of the history of the American Race-Horse to the examination of other races and types of this animal in general use in our country, it must be borne in mind, as before remarked, that the thorough-bred horse of America is the only family of the horse on this continent of pure and unmixed blood.

In the United States, and British America, the process of absorption, or abolition of all the old special breeds, and of the amalgamation of all into one general race, which may fairly be termed specially "American," possessing a very large admixture of thorough blood, has gone on far more rapidly than in England—the rather that, with the one solitary exception of the Norman horse in Canada, no special breeds have ever taken root as such, or been bred, or even attempted to be bred in their purity, in any part of America.

In Canada East, the Norman horse, imported by the early settlers, was bred for many generations entirely unmixed; and, as the general agricultural horse of that province, exists so yet, stunted somewhat in size by the cold climate, and the rough usage to which he has been subjected for centuries, but in no wise degenerated; for he possesses all the honesty, courage, endurance, hardihood, soundness of constitution, and characteristic excellence of feet and legs of his progenitor. Throughout both the provinces, he may be regarded as the basis of the general horse, improved as a working animal by crosses of English half-bred sires; and as a roadster, carriage-horse, a higher class riding or driving horse, by an infusion of English thorough blood.

All these latter types are admirable animals; and it is from the latter admixture that have sprung many of the most celebrated trotting horses, which, originally of Canadian descent, have found their way into the New England States and New York, and there won their laurels as American trotters. Still, it is not to be denied, that there are in different sections of the United States, different local breeds of horses, apparently peculiar, and now become nearly indigenous to those localities, and that those breeds differ not a little, as well in qualities as in form and general appearance. A good judge of horse-flesh, for instance, will find little difficulty in selecting the draught-horse of Boston, that is to say, of Massachusetts and Vermont, from those of New York and New Jersey, or any of the three from the large Pennsylvania team-horses, or from the general stock of the Western States.

The Vermont draught-horse, and the great Pennsylvania horse, known as the Conestoga horse, appear in some considerable degree to merit the title of distinct families; inasmuch as they seem to reproduce themselves continually, and to have done so from a remote period, comparatively speaking, within certain regions of country, which have for many years been furnishing them in considerable numbers to those markets, for which their qualities render them most desirable.

With the limited information at present accessible as to the origin and derivation of these various families, nothing more can be done in the present work than to describe the characteristic points of the breeds in question; and, by comparison with existing foreign races, to approach conjecturally the blood from which they are derived, and also the manner in which they have been originated, where they are now found.

THE VERMONT DRAUGHT-HORSE.

No person familiar with the streets of New York can have failed to notice the magnificent animals, for the most part dark bays, with black legs, manes, and tails, but a few browns, and now and then, but rarely, a deep, rich, glossy chestnut, which draw the heavy wagons of the express companies in that city. They are the very model of what draught-horses should be; combining immense power with great quickness, a very respectable turn of speed, fine show, and good action.

These animals have almost invariably lofty crests, thin withers, and well set-on heads; and, although they are emphatically draught-horses, they have none of that shagginess of mane, tail, and fetlocks, which indicates a descent from the black horse of Lincolnshire, and none of that peculiar curliness or waviness which marks the existence of Canadian or Norman blood for many generations, and which is discoverable in the manes and tails of very many of the Morgan horses.

The peculiar characteristics of these horses are, however, the shortness of their backs, the roundness of their barrels, and the closeness of their ribbing-up. One would say, that they are ponies, until he comes to stand beside them, when he is astonished to find that they are oftener over, than under, sixteen hands in height.

THE VERMONT DRAUGHT-HORSE.

Nine out of ten of these horses are from Vermont; and not only are they the finest animals in all the United States, for the quick draught of heavy loads, but the mares of this stock are incomparably the likeliest, from which, by a well-chosen thorough-bred sire, to raise the most magnificent carriage-horses in the world.

As to the source of this admirable stock of horses, it may be said, that the size, the action, the color, the comparative freedom from hair on the limbs, the straightness of the longer hairs of the mane and tail, and the quickness of movement, would at once lead one to suspect a large cross, perhaps the largest of any, on the original mixed country horse, of Cleveland Bay. There are, however, some points in almost all of these horses, which must be referred to some other foreign cross than the Cleveland, not thorough bred, and certainly, as above remarked, not Norman or Canadian, of which these animals do not exhibit any characteristic. These points are, principally, the shortness of the back, the roundness of the barrel, the closeness of the ribbing-up, the general punchy or pony build of the animal, and its form and size, larger and more massively muscular than those of the Cleveland Bay, yet displaying fully as large, if not a larger, share of blood than belongs to that animal in its unmixed form.

The prevalent colors of this breed also appear to point to an origin different, in part, from that of the pure Cleveland Bays, which lean to the light or yellow bay variation, while these New Englanders tend as decidedly to the blood bay, if not to the brown bay, or pure brown. These latter are especially the dray-horse colors, and the points above specified are those, in a great measure, of the improved dray-horse. The cross of this blood in the present animal, if there be one, is doubtless very remote; and, whether it may have come from a single mixture of the dray stallion long since, or from some half-bred imported stallion, perhaps got by a three-part thorough bred and Clevelander from a dray mare, must, of course, be doubtful. One need have little hesitancy in asserting that the bay draught-horse of Vermont, has in its veins, principally

Cleveland Bay blood, with some cross of thorough blood, one at least, directly or indirectly, of the improved English dray-horse, and not impossibly a chance admixture of the Suffolk.

THE CONESTOGA HORSE.

In appearance this noble draught-horse approaches far more nearly to the improved light-class London dray-horse, and has little, if any, admixture of Cleveland Bay, and certainly none

A CONESTOGA—THE GREAT PENNSYLVANIA DRAUGHT-HORSE.

of thorough blood. He is a teamster, and a teamster only; but a very noble, a very honest, and a moderately quick-working teamster. In size and power some of these great animals employed in draught upon the railroad track in Market street, Philadelphia, are little, if at all, inferior to the dray-horses of the best breweries and distilleries in London; many of them coming fully up to the standard of seventeen or seventeen and a half hands in height.

In color, also, they follow the dray-horses; being more often blood-bays, brown, and dapple-grays than of any other shade. The bays and browns, moreover, are frequently dappled also in their quarters, which is decidedly a dray-horse characteristic and beauty; while it is, in some degree, a derogation to a horse pretending to much blood. This peculiarity is often observable also in the larger of the heavy Vermont draught-horses, and is not unknown in the light and speedy Morgan.

They have the lofty crests, shaggy volumes of mane and tail, round buttocks, hairy fetlocks, and great round feet of the dray-horse; they are, however, longer in the back, finer in the shoulder, looser in the loin, and perhaps, fatter in the side than their English antitypes. They do not run to the unwieldy superfluity of flesh, for which the dray-horse is unfortunately famous; they have a lighter and livelier carriage, a better step and action, and are, in all respects, better travelers, more active, generally useful, and superior animals.

They were for many years, before railroads took a part of the work off their broad and honest backs, the great carriers of produce and provisions from the interior of Pennsylvania to the seaboard, or the market; and the vast white-topped wagons, drawn by superb teams of the stately Conestogas, were a distinctive feature in the landscape of that great agricultural State. The lighter horses of this breed, were the general farm-horses of the country; and no one, who is familiar with the agricultural regions of that fine State, can fail to observe that the farm-horses generally, whether at the plough, or on the road, are of considerably more bulk and bone than those of New York, New Jersey, or the Western country.

Of the Conestoga horse, although it has long been known and distinguished by name as a separate family, nothing is positively authenticated, from the fact that such pedigrees have never been, in the least degree, attended to; and, perhaps, no less from the different language spoken by the German farmers, among whom this stock seems first to have obtained, and by whom principally it has been preserved. It would appear, however, most probable, taking into consideration the thrifty character, and apparently ample means of the early German settlers, their singular adherence to old customs and conservatism of old-country ideas, that they brought with them horses and cattle, such as Wouvermans, and Paul Potter painted; and introduced to the rich pastures of the Delaware and the Schuylkill, the same type of animals which had become famous in the similarly constituted lowlands of Flanders, Guelderland, and the United Provinces.

THE CANADIAN HORSE.

The Canadian is generally low-sized, rarely exceeding fifteen hands, and more often falling short of it. His characteristics are a broad, open forehead; ears somewhat wide apart, and not unfrequently a basin face; the latter, perhaps, a trace of the far remote Spanish blood, said to exist in his veins; the origin of the improved Norman or Percheron stock, being, it is usually believed, a cross of the Spaniard, Barb by descent, with the old Norman war-horse.

His crest is lofty, and his demeanor proud and courageous. His breast is full and broad; his shoulder strong, though somewhat straight, and a little inclined to be heavy; his back broad, and his croup round, fleshy, and muscular. His ribs are not, however, so much arched, nor are they so well closed-up, as his general shape and build would lead one to expect. His legs and feet are admirable; the bone large and flat, and the sinews big, and nervous as steel-springs. His feet seem almost unconscious of disease. His fetlocks are shaggy; his mane voluminous and massive, not seldom, if untrained, falling on both sides of his neck; and his tail abundant; both having a peculiar crimpled wave, never seen in any horse which has not some strain of this blood.

He cannot be called a speedy horse in his pure state; but he is emphatically a quick one, an indefatigable, undaunted traveler, with the greatest endurance, day in and day out, allowing him to go his own pace—say from six to eight miles the hour—with a horse's load behind him, or an animal one can derive. He is extremely hardy, will thrive on any thing, or almost on nothing; is docile, though high-spirited, remarkably sure-footed on the worst ground, and has fine, high action, bending his knee roundly, and setting his foot squarely on the ground. As a farm-horse and ordinary farming roadster, there is no better or more honest animal; and, as one to cross with other breeds, whether upwards by the mares to thorough-bred stallions, or downwards by the stallions to common country mares of other breeds, he has hardly any equal.

From the upward cross, with the English or American thorough-bred on the sire's side, the Canadian has produced some of the fastest trotters and the best gentleman's road and saddle-horses in the country; and, on the other hand, the Canadian stallion, wherever he has been introduced, as he has been largely in the neighborhood of Skaneateles, and generally in the western part of the State of New York, is gaining more and more favor with the farmers, and is improving the style and stamina of the country stock. He is said, although small himself in stature, to have the unusual quality of breeding up in size with larger and loftier mares than himself, and to give the foals his own vigor, pluck, and iron constitution, with the frame and general aspect of their dams. This, it may be remarked in passing, appears to be a characteristic of the Barb blood above all others, and is a strong corroboration of the legend, which attributes to him an early Andalusian strain.

THE INDIAN PONY.

The various breeds of Indian ponies found in the West, generally appear to be the result of a cross between the Southern mustang, descended from the emancipated Spanish horses of the Southwest, and the smallest type of the Canadian, the proportions varying according to the localities in which they are found; those further to the South sharing more largely of the Spanish, and those to the North of the Normal blood.

These little animals, not exceeding thirteen hands in height, have, many of them, all the characteristics of the pure Canadians, and, except in size, are not to be distinguished from them. They have the same bold carriage, open countenance, abundant hair, almost resembling a lion's mane, the same general build, and, above all, the same iron feet and legs. They are merry goers, and over a hard and good road can spin along at nearly nine miles in the hour. They are distinguished for their wonderful sure-footedness, sagacity, and docility. They are driven without blinkers or bearing reins, and where, as is often the case, bridges seem doubtful, the bottom of miry fords suspicious of quagmires, or the road otherwise dangerous, they will put down their heads to examine, try the difficulty with their feet, and, when satisfied, will get through or over places which seem utterly impracticable.

Whence this peculiar pony breed of Canadians has arisen cannot with certainty be traced; it seems, however, to be almost entirely peculiar to the Indian tribes, and, therefore, may have been produced by the dwarfing process, which will arise from hardship and privation, endured for generation after generation, particularly by the young animals and mares while heavy with foal. Most of these animals have no recent cross of the Spanish horse; although some ponies approaching nearly to the same type, show an evident cros of the Mustang; and many animals called Mustangs, have in them some unmistakable Canadian blood.

THE NARRAGANSETT PACER.

This beautiful animal, which, so far as can now be ascertained, has at present entirely ceased to exist, and concerning which the strangest legends and traditions are afloat, was, it may be asserted with comparative certainty, of Andalusian blood. The legends, to which allusion has been made, are two-fold; or, rather, there are two versions of the same legend; one saying that the original stallion, whence the breed originated, was picked up at sea, swimming for his life, no one knew whence or thither, and, that he was so carried in by his salvors to the Providence Plantations; the other, evidently another form of the same story, stating that the same original progenitor was discovered running wild in the woods of Rhode Island.

The question, however, thus far seems to be put at rest by the account of these animals, given in a note to the very curious work, "America Dissected," by the Rev. James McSparran, D. D., which is published as an appendix to the History of the Church of Narragansett, by Wilkins Updike. In this work, the Doctor twice mentions the pacing horse, which was evidently at that remote date, (1721–59,) an established breed in that province. "To remedy this," he says—"this" being the great extent of the parishes in Virginia, of which he is at first speaking, and the distance which had to be traveled to church,—"as the whole province between the mountains, two hundred miles up, and the sea, is all a champaign and without stones, they have plenty of a small sort of horses, the best in the world, like the little Scotch Galloways; and 'tis no extraordinary journey to ride from sixty to seventy miles, or more, in a day. I have often, but upon large *pacing horses*, rode fifty, nay, sixty miles a day, even *here in New England*, where the roads are rough, stony, and uneven." Elsewhere he speaks more pointedly of the same breed. "The produce of this Colony," (Rhode Island,) "is principally butter and cheese, fat cattle, wool, and fine horses, which are exported to all parts of English America. They are remarkable for fleetness and *swift pacing*, and I have seen some of them *pace a mile in a little more than two minutes, and a good deal less than three*."

If the worthy doctor of divinity was a good judge of pace, and had a good timing watch, it would seem that the wonderful me of our fleetest racers was equaled, if not outdone, upwards of a century ago; at all events, he establishes, beyond a peradventure, the existence of the family, and its unequaled powers both of speed and endurance.

To the latter extract is attached a lengthy note, a portion of which we give. "The breed of horses, called 'Narragansett Pacers,' once so celebrated for fleetness, endurance, and speed, has become extinct. These horses were highly valued for the saddle, and transported the rider with great pleasantness and sureness of foot. The pure blood could not trot at all. Formerly, they had pace races. Little Neck Beach, in South Kingston, one mile in length, was the race-course. A silver tankard was the prize, and high bets were otherwise made on the speed. Some of these prize tankards were remaining a few years ago. Traditions respecting the swiftness of these horses are almost incredible. Watson, in his 'Historical Tales of Olden Times,' says: 'In olden time, the horses most valued were *pacers*, now so odious deemed. To this end the breed was propagated with care. The Narragansett pacers were in such repute, that they were sent for, at much trouble and expense, by some who were choice in their selections.'"

The most natural reason assignable for the extinction of this breed, would seem to be somewhat as follows. Up to the beginning of the present century in this country,—much as it was half a century yet farther back in England,—the roads were so bad, as to be, except in the finest weather, utterly impracticable for wheel-carriages; and that, except on the great turnpike-roads, and in the immediate vicinity of the larger towns, private pleasure-vehicles were almost unknown; all long journeys, with few exceptions, all excursions for pleasure or for ordinary business, and all visitings between friends and neighbors being performed by both sexes on the saddle. At that time there was, therefore, a demand, as an actual necessity, for speedy, and, above all, for easy and pleasant-going saddle-horses.

Pacers, whenever they could be found, would most readily answer the desired end.

The expense of this was, of course, considerable, since the pacer could not be used for any other purpose; when, therefore, the roads improved, in proportion to the improvement of the country and the general increase of the population, wheel-carriages generally came into use, and the draught-horse took the place of the saddle-horse. It was soon found that a horse could not be kept even tolerably fit for the saddle, if he was allowed to work in the plough or draw the team, while the same labor in no wise detracted from the chaise or carriage-horse. The pacer, therefore, gave way to the trotter; and the riding-horse, from being an article of necessity, became exclusively one of luxury; to such a degree, that, until comparatively a recent period, when ladies began again to take up riding, there have been very few distinctively broken riding-horses, and still fewer kept exclusively as such in the Northern States of America.

This, unquestionably, is the cause of the extinction of the pacer, although there have been pacing-horses in the eastern section of this country, professedly from Rhode Island, and called by names implying a Narragansett origin; and although it may well be that they were from that region, and possibly, in a remote degree, from that blood, yet they did not pace naturally because they were Narragansett Pacers, but were so called, because coming somewhere from that region of country, they paced by accident—as many chance horses do—or, in some instances, had been taught to pace.

Considering the rare qualities of this variety, and its admirable adaptedness for many purposes of pleasure and convenience, it is a matter for real regret that the family has entirely disappeared, presumably without any prospect or hope of its resuscitation.

THE MORGAN HORSE.

Within a few years past the sporting world have become familiar with a class or type of horses coming from the State of Vermont, known as the Morgan horse; in behalf of which a claim has been made, that it is a distinct family, directly descended from a single horse, owned a little

ETHAN ALLEN—A FAST TROTTING MORGAN HORSE.

before and a little after the commencement of the present century, by Justin Morgan, of Randolph, in Vermont, from whom the name takes its rise.

Without choosing to go into an examination of the validity of this claim—relative to which question an amount of bickering, crimination and recrimination has sprung up, sufficient to furnish the stock in trade of all our stump orators for the next fifty Presidential campaigns—we content ourselves here with alluding to the strong points and excellencies of this particular variety, (for such the most sturdy opponents to its rank as a distinct family freely admit that it possesses,) referring the reader, who is curious in such matters, to the appropriate treatises for and against the claim, which have been as voluminous as the most prolix of Presidential messages.

"The original, a 'Justin Morgan'"—we now quote from "Morgan Horses," by D. C. Linsley—"was about fourteen hands high, and weighed about nine hundred and fifty pounds. His color was dark-bay, with black legs, mane, and tail. He had no white hairs upon him. His mane and tail were coarse and heavy, but not so massive as has been sometimes described; the hair of both was straight, and not inclined to curl. His head was good, not extremely small, but lean and bony, the face straight, forehead broad, ears small and very fine, but set rather wide apart. His eyes were medium size, very dark and prominent, and showed no white around the edge of the lid. His nostrils were very large, the muzzle small, and the lips close and firm. His back and legs were, perhaps, his most noticeable points. The former was very short; the shoulder-blades and thigh-bones being very long and oblique, and the loins exceedingly broad and muscular. His body was rather long, round and deep, close-ribbed up; chest deep and wide, with the breast-bone projecting a good deal in front. His legs were short, close-jointed, thin, but very wide, hard and free from meat, with muscles that were remarkably large for a horse of his size; and this superabundance of muscle manifested itself at every step. His hair was short, and at almost all seasons short and glossy.

He had a little long hair about the fetlocks, and for two or three inches above the fetlock, on the back-side of the legs; the rest of his limbs were entirely free from it. His feet were small, but well-shaped; and he was in every respect perfectly sound and free from blemish. He was a very fast walker. In trotting, his gait was low and smooth, and his step short and nervous; he was not what in these days would be called fast, and we think it doubtful whether he could trot a mile much, if any, within four minutes, although it is claimed by many that he could trot in three.

"Although he raised his feet but little, he never stumbled. His proud, bold, and fearless style of movement, and his vigorous, untiring action have, perhaps, never been surpassed. * * * * * * He was a fleet runner at short distances. Running short distances for small stakes, was very common in Vermont fifty years ago. Eighty rods was very generally the length of the course which usually commenced at a tavern or grocery, and extended the distance agreed upon up or down the public road. In these races the horses were started from a scratch; that is, a mark was drawn across the road in the dirt, and the horses, ranged in a row upon it, went off at the dropping of a hat, or some other signal.

"It will be observed that the form of Justin Morgan was not such as, in our days, is thought best calculated to give the greatest speed for a short distance. Those who believe in long-legged racers will think his legs, body, and stride, were all too short, and to them it may, perhaps, seem surprising that he should be successful, as he invariably was, in such contests."

The qualities claimed for this stock are neat style, good trotting action, great honesty, great quickness and sprightliness of movement,—apart from extraordinary speed, which is not insisted upon as a characteristic of the breed, although some have possessed it—and considerable powers of endurance. There has been some conflict of opinion concerning the courage and endurance of the Morgans, and their ability to maintain a good stroke of speed, say ten miles an hour, for several hours in succession; but it is now well established that this exception has not been fairly taken, and that these horses lack neither courage nor ability to persevere, though not at a high rate of speed.

By fair deduction from the various conflicting accounts of the Morgans, as they now exist, it may be stated that they are a small, compact, active style of horse, showing the evidence of a strain of good blood. They rarely, if ever, exceed fifteen hands two inches, and it is probable that a hand lower, or from that up to fifteen, is nearer to their standard. They are not particularly closely ribbed up, and many of them incline to be sway-backed. Their hind-quarters are generally powerful, and their legs and feet good. There is an evident family resemblance in their foreheads, their neck and crests being so often, as to render the mark somewhat characteristic, lofty but erect, without much curvature, and the neck apt to be thick at the setting-on of the head, which, though good, is rarely blood-like. The manes and tails are almost invariably coarse, as well as heavy and abundant, and have very often a strong wave, or even curl, of the hair.

It is admitted by the most strenuous opponents of this horse as a distinct family, that the very best general stock for breeding for general work—namely, a high cross of the very best thorough-bred on the sires side, with the very best general stock on the dam's—is to be found, so far as the United States are concerned, on the frontiers of Vermont, and that of the most approved quality.

Having given the history of the various types or families of the horse throughout the world, we next propose taking up

THE NATURAL HISTORY OF THE HORSE.

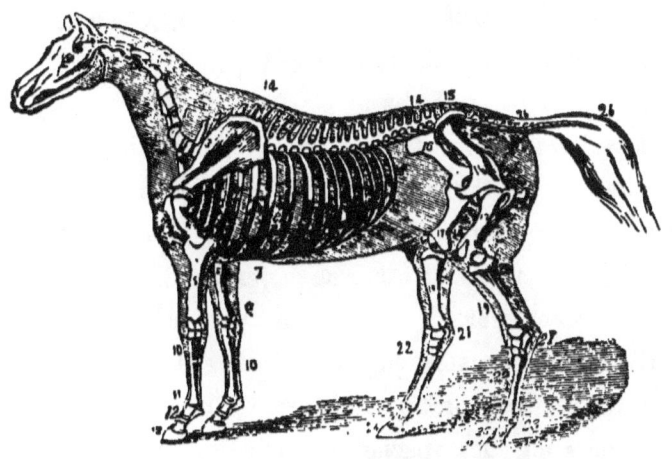

SKELETON OF THE HORSE AS COVERED BY THE MUSCLES.

1, 1. The seven cervical vertebræ, or bones of the neck. 2. The sternum, or breast-bone. 3. The scapula, or shoulder-blade. 4. The humerus, or bone of the arm. 5, 5. The radius, or bone of the fore-arm. 6. The ulna, or elbow. 7. The cartilages of the ribs. 8, 8, 8. The costæ, or ribs. 9. The carpus, or seven bones. 10, 10. The metacarpal, or shank-bones: the larger metacarpal, or cannon, or shank-bone, in front; and the small metacarpal, or splint-bone, behind. 11. The upper pastern. 12. The lower pastern. 13. The coffin-bone. 14 to 14. The eighteen dorsal vertebræ, or bones of the spine. 15. The six lumbar vertebræ, or bones of the loins. 16, 16. The haunch, consisting of the ilium, the ischium, and the pubis. 17, 17. The femur, or thigh-bone. 18, 18. The stifle-joint, with the patella, or knee-cap. 19, 19. The tibia, or proper leg-bone. 20. The fibula. 21, 21. The tarsus, or hock. 22. The metatarsal bones of the hind-leg. 23, 23. The pastern of the hind-feet, including the upper and larger bone, (fig. 23,) the lower pastern, (fig. 25,) and the coffin-bone. (fig. 24.) 26, 26. The caudal vertebræ, or tail-bones.

DIVISION. *Vertebrata*—possessing a back-bone.

CLASS. *Mammalia*—such as give suck.

ORDER. *Pachydermata*—thick-skinned.

FAMILY.   *Solipeda*—uncleft-footed.
GENUS.   *Equus*—the horse family.

With the horse are ranked all those quadrupeds, whose generic distinction is the undivided hoof—the equine genus.

Equus caballus, the horse.

Equus Hemionus, the dziggtai, Asiatic.

Equus Zebra, the zebra, ⎫
Equus Burchelli,   ⎬ South African.
Equus Quagga, the Quagga, ⎭

Equus Asinus, the ass.   0

The horse by far the noblest of the genus, is easily distinguishable from the rest of the group. His varieties are exceedingly numerous, and differ widely in physical appearance. The effects of climate and other agencies are displayed in his frame. It has been asserted, though not upon sufficient basis, that he arrives at the greatest perfection between the fifteenth and fifty-fifth degrees of northern latitude. The mare is found capable of generating her species as early as the second year of her existence; but it is detrimental to her form and the future energies of her offspring, thus prematurely to tax the productive powers of her frame. It would be far more profitable to delay this important function to the fourth or fifth year, when the outline of her form approximates more closely to that of the adult, and the vital energies of the animal economy become more confirmed. Mares, in common with the females of many other quadrupeds, are subject to a periodical appetency for the male, which in them is termed horsing. The natural season of its first occurrence is from the end of March to July, and so providential is this arrangement, that the foal will be produced at a time when nourishment is plentiful for its support.

Common assertion fixes the period of gestation, or the time intervening between conception and foaling, at eleven months; whether lunar or calendar, is not explained. This discrepancy will appear the more unsatisfactory, when it is recollected that eleven calendar months want but two days of twelve lunar months. By various investigations made in France, it has been established that the term of eleven calendar months was often exceeded by several weeks; and sometimes, though less frequently, parturition took place within that period. Some breeders entertain an opinion, that old brood-mares carry the foal considerably longer than young ones; but no satisfactory evidence is offered by them in support of this opinion.

The indications of approaching parturition are enlargement of the external parts of generation, and a gummy exudation from the orifice of the teats. Birth generally takes place within twenty-four hours after the appearance of the latter symptom; but the first acts as a warning, by preceding it for several days. It is but seldom that the mare requires manual assistance at the time of foaling, which generally takes place, without difficulty or danger in the night. The mare, unlike the generality of quadrupeds, foals standing. She rarely produces twins, and when double births do occur, the offspring almost invariably die.

As great facility of motion appears to have been designed by nature in the formation of the horse, many physical peculiarities contribute to insure that end. A bulky, pendulous udder, like that possessed by some of the *ruminantia*, would be incompatible with that quality. The *mamma*, therefore, is small, and furnished with only two teats, which supply a milk of a highly nutritious character, and possessing a larger quantity of saccharine matter than any other animal is known to possess.

The disproportionate length of the foal's legs, which is so strikingly apparent, when compared with those of the adult animal, is thought by some naturalists to be provided by prescient nature to enable the young animal to keep pace with its dam during flight from any menacing danger. Linnaeus attempted to ascertain the future height of the colt by admeasurement of its legs; but so much is found to depend upon the quantity and character of the nutriment with which it is provided during the period of its growth, that little reliance can be placed upon early experiments of this kind. The historian, the warrior, and the horseman, Xenophon, has long ago alluded to the same subject in his treatise upon horsemanship. "I now explain," said he, "how a man may run the least risk of being deceived, when conjecturing the future height of a horse. The young horse, which, when foaled, has the shank-bone the longest, invariably turns out the largest. For, as time advances, the shank-bones of all quadrupeds increase but little; but that the rest of the body may be symmetrical, it increases in proportion."

Puberty commences in both sexes as early as the second year, but all the structures continue to be gradually developed until the end of the fifth year, by which time the changes in the teeth are perfected, and the muscles have acquired a growth and tone which give to the form the distinctives of adolescence. It is during the term which elapses between the period of adult age and that of confirmed virility, that a further progressive change takes place in the animal economy; the powers of the whole frame continue to acquire strength, and although there is no further increase in height, the *parietes* of the large cavities and the muscles of voluntary motion assume a finished and rotund appearance, and render the animal more capable of enduring continued privation and exertion; the vital endurance and resistance being greater than during the period of adolescence. The fire and expression of the head, the spirit, character, and disposition, become also more marked toward the termination of this epoch.

The natural period of the decay of the vital powers, senility, and mature death, may be conjectured to be about thirty years; but few horses, owing to our barbarous treatment, attain that term.

The walk, the trot, and the gallop are the usually well-known natural paces of the horse; but the fact of some individuals contracting the pace called amble (which is neither racking or pacing), without previous tuition, has induced many writers to regard that also as a natural method of progression.

In England, and other northern countries, on the approach of mild weather, the horse, by a natural process, analogous to moulting in birds, divests himself of his winter's clothing of long hair, and produces one of a shorter and cooler texture; and again, before the recurrence of cold weather, reassumes his warm and lengthened coat to protect himself from the inclemency of the approaching season. The autumnal change is not by any means so general as that which takes place at the commencement of spring; in America, however, at least in the northern parts, this change is invariable. The hair is not so completely changed; only a portion of it is thrown off, and that which remains, with that which springs up, grows long, and is adapted to the temperature of the atmosphere. These alternate changes are not so well marked in countries possessing an even temperature, nor even are they so plainly seen in horses kept in the warm atmosphere of a close stable all the year round. When the shedding of the coat commences, the bulbs of the old hair become pale, and by the side of each a small black globular body is formed, which is soon developed into the new hair. Thus the *matrix* of the new hair is not the old bulb, but it is based in another productive follicle. The long hair of the mane, tail, and fetlocks is not shed at definite periods with that of the body, but is replaced by a shorter and more uniform process. The hair of the mane and tail will, if protected, grow to an almost incredible length.

The property of changing the color of the hair with the season, possessed by many animals of the arctic region, adapting them to the temperature, is also manifested in the horse, though in a much less degree, for it may be seen that when constantly exposed to the elements, the long winter-coat assumes a much lighter hue than that of its predecessor.

The horse in common with many other animals, is provided with a thin, sub-cuticular muscle, covering the shoulders, flanks, and sides, whose use is to corrugate the skin, shake off flies, and dislodge other annoying substances.

The sense of smell is so delicately acute in the horse, that perhaps he is not exceeded in this function by any other animal. The nose is provided with a very extensive surface for the distribution of the olfactory nerve, by the curious foldings of the turbinated bones. It is principally by means of this faculty that he is enabled to distinguish the qualities of the plants upon which he feeds, and to reject such as are of a noxious or poisonous description. "Nature," said Linnæus, "teaches the brute creation to distinguish, without a preceptor, the useful from the hurtful, while man is left to his own inquiries." On putting the finger into the nostrils, at the upper and outward parts, they pass into blind pouches of considerable dimensions. These curious cavities have nothing to do with smelling, because they are lined with a reflection of common integument, but they may possibly be of use in mechanically distending the external entrance of the nostrils, and thus materially facilitate respiration during violent exertion. They are also brought into use when the animal neighs; and the Hungarian soldiery slit them up, to preclude the possibility of being prematurely discovered to the enemy by the exercise of this habit. It is worthy of remark, in this connection, that the preference of Arabs for the mare to the horse, for warlike purposes, is attributable to the fact that they do not neigh when they scent the vicinity of other horses, as stallions invariably do—the Arabs never attacking, save by surprise. Those nations which fight by open force have no such preference, but mainly use the stallion. On the lower part of the nostril, toward the outer edge, may be seen the mouth of a small tube, which conveys the tears from the inner *canthus*, or corner of the eye. It opens on the skin just before it joins the lining membrane of the nose. This little cavity has often been mistaken, by unqualified persons, for an ulcer common in glanderous affections, and the poor animal has frequently fallen a victim to the error.

Their eyes are large in proportion to those of some other quadrupeds, and the pupilar opening is of an oblate elliptic form, with its long axes parallel to the horizon, thus increasing the lateral field of vision. Round the edges of the pupil is a curious fringe of deep plum-colored eminences, supposed to be serviceable in absorbing the superabundant rays of light which may be transmitted to the eye. The horse's sight is excellent, and, although not regarded as a nocturnal animal, he can distinguish objects at night with great facility. There are but few horsemen, who have not benefited by this power, when the shades of night have fallen round them.

The ears are comparatively small, but the conch is endowed with extensive motion, so as to catch the sound coming from

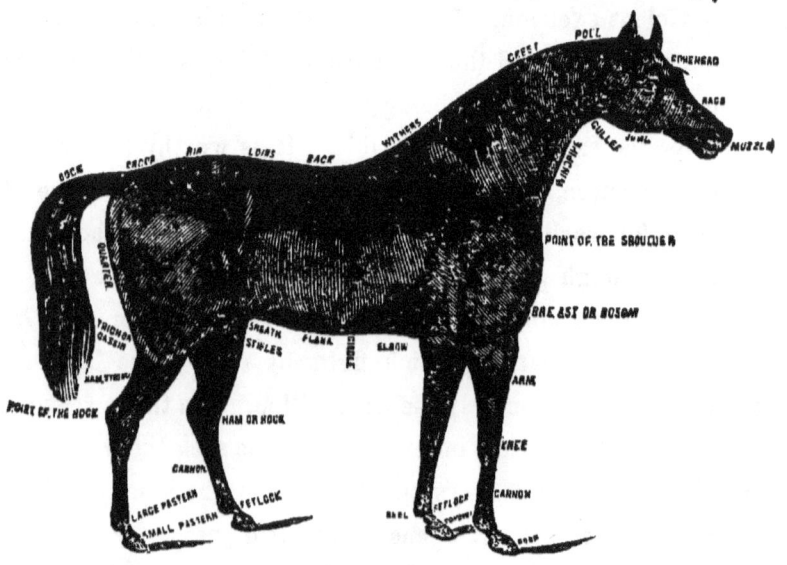

NAMES APPLIED TO THE VARIOUS EXTERNAL PARTS OF THE HORSE.

any quarter. Their hearing is quick, and although blindness is so distinctively prevalent among horses, deafness is exceedingly uncommon. During sleep, one ear is usually directed forward, and the other backward; when on a march at night, in company, it has been noticed "that those in the front direct their ears forward, those in the rear backward, and those in the centre turn them laterally, or across; the whole troop seeming thus to be actuated by one feeling, which watches the general safety." In contests of speed the ears are generally laid backwards, so as to afford no opposition to the rapid progress of the animal. It must be evident, that if the concave surfaces of these organs be presented forward, they would gather the wind, and slightly impede progression. Another reason assigned for this deflection, is that the animal may avoid the inconvenience, if not pain, which the current of air, produced by his velocity, would inflict on that delicate organ.

The different vocal articulations to which the horse gives utterance, are collectively termed neighing; but some variety of intonation may be discovered in the expression of its passions; as, for instance, the cry of joy or recognition differs in a marked manner from that of desire; and anger from playfulness. The females do not neigh so frequently, nor with so much force as the males. Castration is known to have a modulating effect upon the voice. It is supposed that neighing is produced by the vibration of two small triangular membranes found at the extremity of the *glottis*. In the ass and mule, this structure is wanting; but they are provided with a more singular apparatus. Hollowed out of the thyroid cartilages is a small concavity, over which is stretched a membrane, similar to the parchment on the head of a drum. When air is forced behind this substance, a very considerable noise is produced, though from the absence of muscularity the vibrations are without modulation, and consequently dissonant.

The intellectual character of the horse is scarcely excelled by that of any other quadruped. His perceptions are remarkably clear, and his memory excellent. Attachment to those who tend him with kindness, forms a prominent trait in his character. The feats which he is often taught to perform, evince a high intellectual capacity. Travelers in the desert assure us that horses possess the faculty of directing their course to the nearest water, when hard pressed for that necessary article.

Horses swim with the greatest facility, and the distances they have been known to perform in the water exceed our expectation. A horse that was wrecked off the coast of South America swam seven miles to land, thus saving his life.

There exist some important differences in the animal economy of the equine family and that of other herbivorous animals, which, as the inferences from them are of some consequence, it is necessary briefly to notice. The horse naturally requires but little sleep, and even that it often takes standing. In a state of nature, when fodder is short, to support itself properly it is compelled to graze twenty hours out of the twenty-four. Ruminating animals eat with greater rapidity, and lie down to chew the cud. The horse eats no faster than it digests. Digestion in the former is interrupted; in the latter, continuous. This explains why the horse has no gall-bladder, as it requires no reservoir for that necessary fluid; for, as fast as the bile is secreted by the liver it is carried to the intestines to perform its important action on the chymous mass. The stomach of the horse is also remarkably small and simple, differing widely from the capacious and complicated structures of the *ruminantia;* but the intestines are long, and the *cæcum* capable of containing a large quantity of fluid, of which it is considered the receptacle. The *mamma* of the mare is by no means so pendulous and bulky as that of the cow. The horse's feet, from their compact, undivided nature, are much less liable to injury during fleet exertion than those of the ox. All these circumstances tend to establish the individuality of the horse, and are so many proofs of admirable design for the purposes to which man has applied him; for, without these peculiarities, he would not be so valuable and superior, as a beast of continued and rapid motion, and would consequently occupy a very inferior station.

Linnæus asserted that the male horse was without the rudimentary *mamma* invariably found in the males of other animals; but this naturalist was mistaken, for they may be seen on each side of the sheath, and, although of no possible use, still their existence preserves the uniformity of nature's operations.

The horse and zebra possess horny callosities on the inside of the fore-legs, above the knees, and on the hocks of the hind-legs; the ass and the quagga have them only on the fore extremities.

In a state of nature, the horse is purely a herbivorous animal, but under the restraint which domestication imposes, his habits become changed, and grain and dry grasses form the principal articles of his diet. Domestication is known to originate many diseases totally unknown in a natural state, but it appears to have the effect of augmenting the muscular power of the animal far beyond its uncultivated state.

It may be remarked, in addition to what has been previously said as to the limit of life allotted to the horse, that there is some difficulty in estimating the natural average length of his life, since many obstacles oppose an inquiry on a scale of sufficient magnitude to be satisfactory. The numerous evils entailed on him by the arduous labors and the r stricted and unnatural habits of a domesticated state tend greatly to abbreviate life. From these and other reasons. it cannot be much doubted that his age is greatly underrated. Horses are most erroneously termed aged on the obliteration of the mark from the lower incisor teeth, which occurs by the completion of the eighth year; and though it is far from being the natural term of age and debility, or even of the decline of the vital energies, it too frequently happens, that by that time bodily infirmities have been prematurely induced by over-exertion of their powers. Horses at twenty years of age, are often met with in cases where the least humanity has been bestowed on their management. Eclipse died at the age of twenty-five; Flying Childers, at twenty-six. Brom's mare Maggie reached more than twenty-nine years. Bucephalus, the celebrated horse of Alexander of Macedon, lived till thirty. The natural age is probably between twenty-five and thirty. A faint and uncertain guide is found in the register of the ages of the most celebrated racing stallions, recollecting, however, that several of them were destroyed on becoming useless for the purposes of the turf. The united ages of ninety-three of these horses amounted to two thousand and five years; or rather more than twenty-one and a half years to each horse.

As a matter of civil economy, it is important to judge correctly of the age of the horse. This is chiefly accomplished by observing the natural changes which occur in his teeth, the periods at which they appear, are shed and replaced, and the alterations in their form and markings.

The teeth of most animals offer some criterion by which their age can be estimated with more or less accuracy. The teeth are nearly the sole indices of the age of the horse, ass, elephant, camel, dog, and the polled varieties of the ox and sheep; while in other domesticated animals, as the elk, deer, goat, common ox and sheep), the horns also present legible indications of the progress of time.

Reference to the teeth to ascertain the age of the horse is not by any means of recent origin. Xenophon, in his work on horsemanship, from which we have already quoted, alludes to it as an established custom used in the selection of cavalry for the Grecian armies; he properly advised the rejection of such horses as have lost the dental mark. The same facts are subsequently noticed by Varro, Columella, Vegetius, and other Roman writers.

The horse, when full-mouthed, possesses forty teeth—twenty in each jaw. They are named from their use, position, and character. Those in the front of the mouth, whose office it is to gather food when grazing, are termed *incisors*, or, more properly, *nippers*. They are twelve in number; six above, and six below. They do not overlap each other, as is the case in man, but meet in a broad tabular surface. From these teeth the age of the animal is principally deduced. For the sake of description, they are usually ranged in pairs, as they appear; and the first pair is called the *central*, the second the *dividers*, and the third the *corner nippers*. The *tushes*, or *canines*, come next; one above, and one below on each side. They are of a pointed form, and are convex on the outer sides, and slightly concave on the inner surface. They scarcely ever appear above the gums in mares, although their rudiments may be discovered on dissection, imbedded in the maxillary bones. They are consequently regarded as sexual distinctions. It is difficult to assign their use; their position precludes the possibility of their being used as weapons of offense or defense. They may be viewed as a link of uniformity so commonly traced in the animated world. The *grinders*, or *molars*, are twenty-four in number. They are teeth of great power. By them the food is crushed or ground into small particles, and prepared for the digestive action of the stomach. In order to fit them for this office, they possess additional interlayers of enamel, which prevent their too rapid wear.

In common with most animals, the horse is provided with two sets of teeth; those appearing first are known as the *temporary, deciduous,* or *milk* teeth, and are succeeded by the *permanent set.* On comparing the different magnitudes of the jaw-bones of the colt and the adult horse, the necessity of such a change is at once apparent. By it the teeth are adapted to the size of the maxillary bones. The teeth, from their peculiar character and mode of growth, do not admit of any material increase of dimension; and nature was therefore forced either to place the large permanent teeth in small and disproportionate jaw-bones, or to adapt the size of the teeth by displacement to the growth of the bones that contained them. The latter process is adopted, and constitutes one of those remarkable evidences of creative power, with which the living frame is replete.

Three substances enter into the structure of the teeth; first the *enamel;* secondly, the *dental bone,* or *ivory;* and thirdly, a *cortical envelope,* surrounding the fang. The enamel differs but little in chemical constitution from the bony body of the teeth; and that principally results from the absence of animal matter in it. It appears closely analogous to the univalve porcellaneous shells, and is the hardest and most indestructible substance of the body. The dental bone is distinctly tubular in structure; these tubuli taking a perpendicular direction, being exceedingly small, but capable of absorbing ink by capillary attraction. No such tubuli have been traced in the enamel. The teeth, both incisors and grinders, are being constantly worn away at the crown; but the loss is supplied by the gradual, continuous, and equivalent growth from the root. The horse's teeth are sometimes, but not frequently, subject to disease. It is seldom that any of them are lost from age, as is the case with man, and most other animals.

It has been remarked, that the constitution of horses and men may be considered as in an equal degree of perfection and capability of exertion, or of debility and decay, according as youth or age preponderates. Thus, the first five years of a horse may be considered as equivalent to the first twenty in man; or thus, that a horse five years old may be comparatively considered as a man of twenty; a horse of ten years, as a man of forty; a horse of fifteen, as a man of fifty; a horse of twenty, as a man of sixty; of twenty-five, as a man of seventy; of thirty, as a man of eighty; of thirty-five, as a man of ninety. So far from this comparison being in favor of the horse, it may rather be regarded as too little. Horses of thirty-five years of age are as common as men of ninety, provided it be taken into account that there are twenty human subjects for every horse; and, unquestionably, a horse of forty-five is less rare than a man of one hundred and ten.

To this it may be added, that the early English racers appear to have been more addicted to longevity than those of modern days, and the American horse generally than the English; probably because, in the former case, the horse was not put to hard work until his powers were developed by an advance toward maturity. Two and three year old training was unknown until a recent date; and, in the latter case, in America horses are little used in harness, or for general work, until they have attained to five or six years.

We will next consider the first appearance and successive changes of the teeth, with the marks and their descriptions from commencement to maturity.

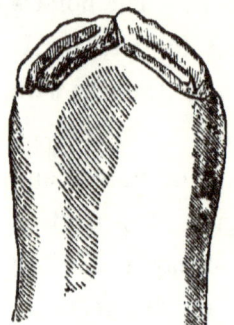
EIGHT DAYS' TEETH.

Seven or eight months before the foal is born, the germs or beginnings of the teeth are visible in the cavities of the jaws. At the time of birth, the first and second grinders have appeared, large, compared with the size of the jaw, seemingly filling it. In the course of seven or eight days, the two centre nippers are seen as here represented.

In the course of the first month, the third grinder appears, above and below; and not long after, and generally before six weeks have expired, another incisor above and below will be seen on each side of the two first, which have now considerably grown, but not attained their perfect height. This cut will then represent the appearance of the mouth.

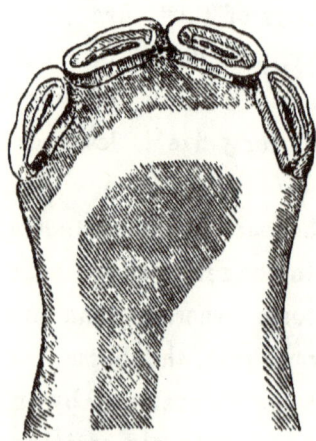
THREE OR FOUR MONTHS' TEETH.

At two months, the centre nippers will have reached their natural level, and between the second and third month the second pair will have overtaken them. They will then begin to wear a little, and the outer edge, which was at first somewhat raised and sharp, is brought to a level with the inner edge, and so the mouth continues until some time between the sixth and ninth month, when another nipper begins to appear on each side of the first two, making six above and below, and completing the colt's mouth; after which the only observable difference, until between the second and third year, is in the wear and tear of these teeth.

These teeth are covered with a polished and exceedingly hard enamel; indeed, it is so hard that it almost bids defiance to the action of a file. It spreads over that portion of the tooth which appears above the gum, and not only so, but as they are to be so much employed in nipping up the grass and gathering the animal's food—and in such employment even this hard substance must be gradually worn away—a portion of it, as it passes over the upper surface of the teeth, is bent inward, and sunk into the body of the teeth, and forms a little pit in them. The inside and bottom of this pit being blackened by the food, constitute the *mark* in them, by the gradual disappearance of which, in consequence of the wearing down of the teeth, we are enabled for several years to judge of the age of the animal.

The colt's nipping teeth are rounded in front, somewhat hollow toward the mouth, and presenting a cutting surface, with the outer edge rising in a slanting direction above the inner edge. This, however, soon begins to wear down, until both surfaces are level, and the mark, which was originally long and narrow, becomes shorter, wider, and fainter. At six months, the four nippers are beginning to wear to a level.

The annexed cut will convey some idea of the appearance of the teeth at twelve months. The four middle teeth are almost level, and the corners are becoming so. The mark in the two middle teeth is wide and faint, in the next two teeth it is longer, darker, and more narrow. In the corner teeth it is longest, darkest, and most narrow.

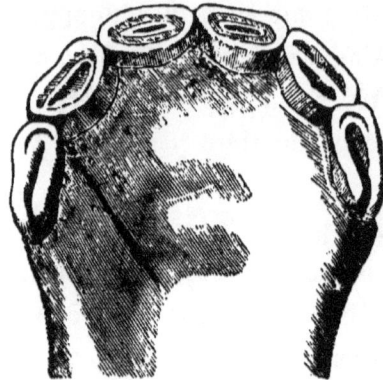
TEETH AT TWELVE MONTHS.

The back teeth, or grinders, will not guide us far in ascertaining the age of the animal, for we cannot easily inspect them; but there are some interesting particulars connected with them. The foal is born with two grinders in each jaw, above and below, or they appear within two or three days after birth. Before the expiration of the month they are succeeded by a third, more backward. The crowns of the grinders are entirely covered with enamel on the tops and sides, but attrition soon wears it away from the top, and there remains a compound surface of alternate layers of crusta petrosa, enamel, and ivory, which are employed in grinding down the hardest portions of the food. Nature has, therefore, made an additional provision for their strength and endurance The annexed cut represents a grinder sawed across. The five dark spots represent bony matter; the parts covered with lines enamel, and the white spaces a strong bony cement uniting the other portions of the teeth.

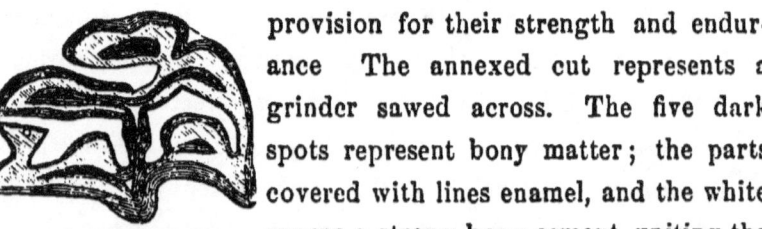
A GRINDER SAWED ACROSS.

At the completion of the first year a fourth grinder usually comes up, and the yearling has then, or soon afterwards, six nippers and four grinders above and below in each jaw, which, with the alteration in the nippers just described, will enable us to calculate the age of foal, subject to some variations, arising from the period of weaning and the nature of the food.

At the age of one year and a half, the mark in the central nippers will be much shorter and fainter; that in the two other pairs will have undergone an evident change, and all the nippers will be flat. At two years this will be more manifest. The accompanying cut deserves attention, as giving an accurate representation of the nippers in the lower jaw of a two-year-old colt.

TWO YEARS TEETH.

About this period a fifth grinder will appear, and now likewise commences another process. The first teeth are adapted to the size and wants of the young animal. They are sufficiently large to occupy and fill the colt's jaws; but when these bones have expanded with the increasing growth of the animal, the teeth are separated too far from each other to be useful, and another and larger set is required. The second teeth then begin to push up from below, and the fangs of the first are absorbed, until the former approach the surface of the gum, when they drop out. Where the temporary teeth do not rise immediately under the milk teeth, but by their sides, the latter being pressed sideways are absorbed throughout their whole length. They grow narrow, are pushed out of place, and cause inconvenience to the gum, and sometimes to the cheek. They are then sometimes improperly called *wolf's* teeth, and should be extracted.

The teeth which first appeared are first renewed, and therefore the front or first grinders are changed at the age of two years. During the period between the falling out of the central milk teeth, and the coming up of the permanent ones, the colt, having a broken mouth, may find some difficulty in grazing. If he should fall away considerably in condition, he should be fed with mashes and corn, or cut feed. The cut annexed represents a three-year-old mouth.

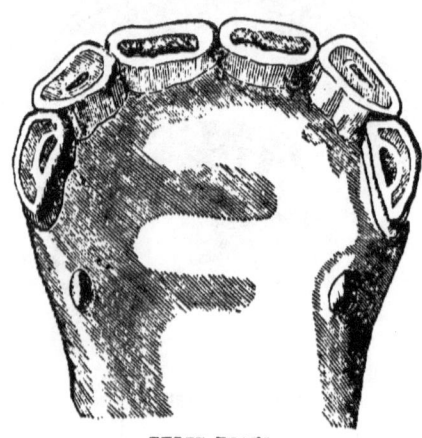

THREE YEARS.

The central teeth are larger than the others, with two grooves in the entire convex surface, and the mark is long, narrow, deep, and black. Not having yet attained their full growth, they are lower than the others. The mark in the next two nippers is nearly worn out, and it is wearing away in the corner nippers.

Is it possible to give this mouth to an early two-year-old?

The ages of all horses used to be reckoned from the first of May; but some are foaled even as early as January, and being actually four months over the two years, if they have been well nursed and fed, and are strong and large, they may, with the inexperienced, have an additional year put upon them. The central nippers are punched or drawn out, and the others appear three or four months earlier than they otherwise would. In the natural process they would only rise by long pressing upon the first teeth, and causing their absorption. But, opposition from the first set being removed, it is easy to imagine that their progress will be more rapid. Three or four months will be gained in the appearance of these teeth, and these three or four months will enable the breeder to term him a late colt of the preceding year. To him, however, who is accustomed to horses, the general form of the animal, the little development of the forehand, the continuance of the mark upon the next pair of nippers, its more evident existence in the corner ones, some enlargement or irregularity about the gums from the violence used in forcing out the teeth, the small growth of the first and fifth grinders, and the non-appearance of the sixth grinder, which, if it be not through the gum at three years old, is swelling under it, and preparing to get through—any or all of these circumstances, carefully attended to, will be a sufficient security against deception.

A horse at three years old ought to have the central permanent nippers growing, the other two pairs wasting, six grinders in each jaw, above and below, the first and fifth level, the others and the sixth protruding. The sharp edge of new incisors, although it could not well be expressed in the cut, will be very evident when compared with the old teeth.

As the permanent nippers wear and continue to grow, a narrow portion of the cone-shaped tooth is exposed by the attrition, and they look as if they had been compressed, but it is not so. Not only will the mark be wearing out, but the crowns of the teeth will be sensibly smaller.

At three years and a half, or between that and four, the next pair of nippers will be changed, and the mouth at that time cannot be mistaken. The central nippers will have attained nearly their full growth. A vacuity will be left where the second stood, or they will begin to peep above the gum, and the corner ones will be diminished in breadth, worn down, and the mark becoming small and faint. At this period, likewise, the second pair of grinders will be shed. Previously to this may be the attempt of the dealer to give to his three-year-old an additional year; but the fraud will be detected by an examination similar to that already described.

At four years, the central nippers will be fully developed; the sharp edge somewhat worn off, and the mark shorter, wider, and fainter. The next pair will be up, but they will be small, with the mark deep and extending quite across them as in the annexed cut. The corner nippers will be larger than the inside ones, yet smaller than they were, and flat, and the mark nearly effaced. The sixth grinders will have risen to a level with the others, and the tushes will begin to appear.

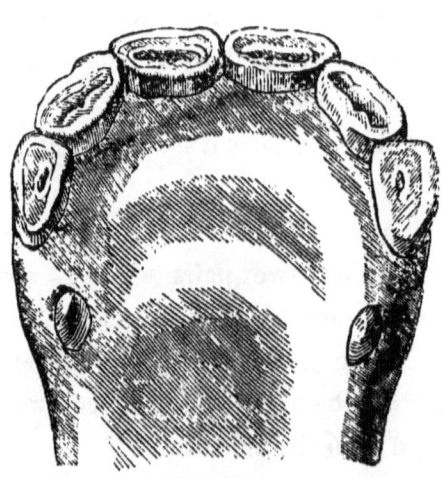

FOUR YEARS.

Now, more than at any other time, will the dealer be anxious to put an additional year upon the animal, for the difference between a four-year-old colt and a five-year-old horse, in strength, utility, and value, is very great; but the want of wear in the other nippers, the small size of the corner ones, the little growth of the tush, the smallness of the second grinder, the low forehand, the legginess of the colt, and the thickness and little depth of the mouth, will, to a man of common experience among horses, at once detect the cheat.

The tushes are four in number, two in each jaw, situated between the nippers and the grinders, much nearer to the former than the latter, and nearer in the lower jaw than in the upper, but this distance increases in both jaws with the age. In shape, the tush somewhat resembles a cone; it protrudes from the gum about half an inch, and is sharp-pointed and curved. The appearance of this tush in the horse may vary from four years to four years and six months. It can only be accelerated a few weeks by cutting the gum over it. At four years and a half, or between that and five, the last important change takes place in the mouth of the horse. The corner nippers are shed, and the permanent ones begin to appear. The central nippers are considerably worn, and the next pair are commencing to show marks of usage. The tush has now protruded, and is generally a full half inch in height; externally, it has a rounded prominence, with a groove on either side, and it is evidently hollowed within. The reader scarcely needs to be told that after the rising of the corner nipper, the animal changes its name. The colt becomes a horse, the filly a mare.

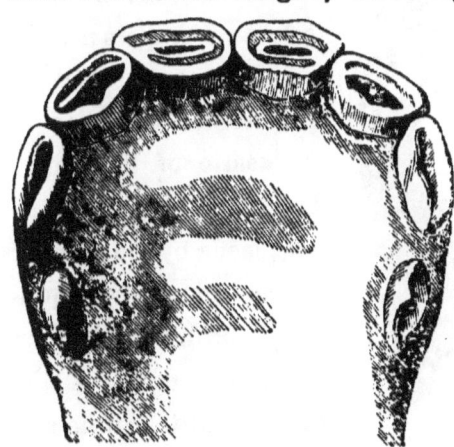

FIVE YEARS.

At five years, the horse's mouth is almost perfect, as represented in the annexed cut. The corner nippers are quite up, with the long, deep mark irregular in the inside, and the other nippers bearing evident tokens of increased wearing. The tush is much grown; the grooves have almost or quite disappeared, and the outer surface is regularly convex. It is still as concave within, and with the edge nearly as sharp, as it was six months before. The sixth molar is quite up, and the third molar is wanting. This last circumstance, if the general appearance of the animal, and particularly his forehand, and the wearing of the centre nippers, and the growth and shape of the tushes be likewise carefully attended to, will prevent deception, if a late four-year-old is attempted to be substituted for a five-year-old. The nippers may be brought up a few months before their time, and the tushes a few weeks, but the grinder is with difficulty displaced. The last three grinders and the tushes are never shed.

SIX YEARS.

At six years, as in the annexed cut, the *mark* on the central nippers is worn out. There will still be a difference of color in the centre of the tooth. The cement filling up the hole, made by the dipping of the enamel, will present a browner hue than the other parts of the tooth; and it will be evidently surrounded by an edge of enamel, and there will remain ever a little depression in the centre, and also a depression round the case of enamel; but the deep hole in the centre of the teeth, with the blackened surface which it presents, and the elevated edge of enamel, will have disappeared. Persons not much accustomed to horses have been puzzled here. They expected to find a plain surface of uniform color, and knew not what conclusion to draw when there were both discoloration and irregularity.

In the next incisors, the mark is shorter, broader, and fainter, and in the corner teeth the edges of the enamel are more regular, and the surface is evidently worn. The tush has attained its full growth, being nearly or quite an inch long, convex outward, concave within, tending to a point, and the extremity somewhat curved. The third grinder is fairly up, and all the grinders are level.

The horse may now be said to have a perfect mouth. All the teeth are produced, fully grown, and have sustained no material injury. During these important changes of the teeth, the animal has suffered less than could be supposed possible.

At seven years, as in the accompanying cut, the mark, in the way in which it has been described, is worn out in the four central nippers, and is fast wearing away in the corner teeth; the tush is also beginning to be altered. It is rounded at the point, rounded at the edges, still round without, and beginning to get round inside.

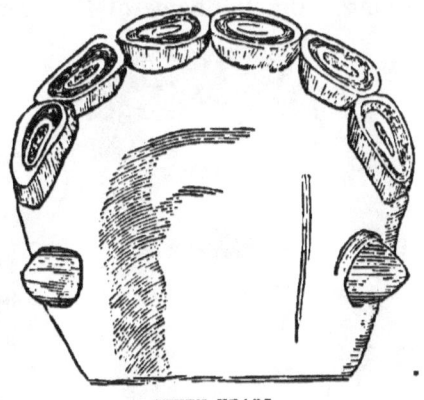
SEVEN YEARS.

At eight years old, the tush is rounder in every way; the mark is gone from all the bottom nippers, and it may almost be said to be out of the mouth. There is nothing remaining in the bottom nippers that can clearly show the age of the horse, or justify the most experienced examiner in giving a positive opinion. This should be distinctly borne in mind, as it is a very common error in the United States, and one especially insisted on by dealers having old horses to sell, that the age can be positively ascertained even to ten, eleven, or twelve years, so that it can be predicated of a horse that he is so old, and no older. This is an absolute fallacy. It is easy, from many general signs, to see that a horse is above eight years old; but it is impossible to judge certainly how much older. The length and angularity of the nippers, the depth of the super-orbital cavities, and other points of information, may enable a good judge to guess comparatively, but never to speak surely.

Dishonest dealers have been said to resort to a method of prolonging the mark on the lower nippers. It is called Bishoping, from the name of the scoundrel who invented it. The horse of eight or nine years old—whose mouth is represented in the accompanying cut—is thrown, and with an engraver's tool a hole is dug in the now almost plain surface of the corner teeth, in shape resembling the mark yet left in those of a seven-year-old horse. The hole is then burned with a heated iron, and a permanent black stain is left. The next pair of nippers is

EIGHT OR NINE YEARS.

sometimes slightly touched. An ignorant man would be very easily deceived by this trick; but the irregular appearance of the cavity, the diffusion of the black stain around the tushes, the sharpened edges and concave inner surface of which can never be given again, the marks on the upper nippers, together with the general conformation of the horse, can never deceive the careful examiner.

Horsemen, after the animal is eight years old, are accustomed to look to the nippers in the upper jaw, and some conclusion has been drawn from the appearances which they present. It cannot be doubted that the mark remains in them for some years after it has been obliterated in the nippers of the lower jaw.

There are various opinions as to the intervals between the disappearance of the mark from the different cutting teeth of the upper jaw. Some have averaged it at two years, others at one. The latter opinion is more commonly adopted by those most conversant, and then the age is thus determined. At nine years, the mark will be worn from the middle nippers; from the next pair at ten; and from all the upper nippers at eleven. During these periods the tush is likewise undergoing a manifest change. It is blunter, shorter, and rounder. In what degree this takes place in the different periods, long and favorable opportunities can alone enable the horseman to decide.

The alteration in the form of the tushes is frequently uncertain. It will sometimes be blunt at eight; and at others remain pointed at eighteen.

After eleven, and until the horse is very old, the age may be guessed at with some degree of confidence, from the shape of the upper surface or extremity of the nippers. At eight they are all oval, the length of the oval running across from tooth to tooth; but as the horse gets older, the teeth diminish in size—and this commencing in their width, and not in their thickness. They become a little apart from each other, and their surfaces become round instead of oval. At nine, the centre nippers are evidently so; at ten, the others begin to have their ovals shortened. At eleven, the second pair of nippers is quite rounded; and at thirteen, the corner ones have also that appearance. At fourteen, the faces of the central nippers become somewhat triangular. At seventeen, they are all so. At nineteen, the angles begin to wear off, and the central teeth are again oval, but in a reversed direction, viz., from outward, inward; and at twenty-one, they all wear this form.

It would, of course, be folly to expect any thing like a certainty in an opinion of the exact age of an old horse, as drawn from the above indications. It is contended by some, though denied by others, that stabled horses have the marks sooner worn out than those that are at grass; and crib-biters still sooner. At nine or ten, the bars of the mouth become less prominent, and their regular diminution will designate increasing age. At eleven or twelve, the lower nippers change their original upright direction, and project forward horizontally, becoming of a yellow color.

The general indications of old age, independent of the teeth, are the deepening of the hollows over the eyes; gray hairs, and particularly over the eyes, and about the muzzle; thinness and hanging down of the lips; sharpness of the withers, sinking of the back, lengthening of the quarters; and the disappearance of windgalls, spavins, and tumors of every kind.

Horses kindly and not prematurely used, sometimes live to between thirty-five and forty-five years of age; and a well authenticated account is given of a barge horse that died in his sixty-second year.

Under this head of age, nothing beyond the cut of the complete aged mouth, with the accompanying description of it, would have been here inserted, were it not for the prevalent opinion, inculcated by interested dealers in the United States, that the age of a horse, after eight or nine years, can be as certainly and as exactly predicated by mouth-mark, and his exact age guaranteed accordingly, as previously to that period.

Summing up all that need be offered on this particular point, we simply say, that if one chooses to buy a horse past mark of mouth, he must do so on his own judgment and at his own risk; for to credit any assertions, or to give ear to any horse-dealer's opinion on the subject, is sheer folly.

MARE AND FOAL

BREEDING AND MANAGEMENT

RELATIVE TO BREEDING,—a very important subject, all will admit—two very common mistakes are made; the first, that mares are bred from only because they are useless for work, and consequently have to be turned out to grass for the season; the second, that a mare is put to a handsome horse which may chance to strike the fancy of her owner, without a moment's consideration on the part of the latter as to how far his particular mare is suited to that particular horse. The consequence of the first error is, that the infirmities of the mare are perpetuated in her unfortunate offspring, and thus become hereditary, to the no small disappointment of the breeder. In the second case mentioned, the result is an indescribable mongrel, possessing only a combination of bad qualities, without a single redeeming trait.

Now, no principle is better established in breeding than that "like will produce like;" in other words, that the offspring will inherit the general or mingled qualities of the parents. So true is this, that there is scarcely a disease affecting either of the parents that is not inherited by the foal, or, at least, to which he does not at times show a predisposition. The consequences of bad usage or hard work even will descend to the progeny. Though the defects may not appear in the immediate offspring, they often do in the next, or some succeeding generation. Some knowledge is therefore indispensable of the parentage both of the sire and the dam.

Both parents should be selected with reference not only to their individual points of excellence, but also to the relative adaptation which the points of one present to the points of the other. Though both may be excellent in their way, one parent may have points of excellence which actually counteract or neutralize those of the other. None but sound parents, therefore, should be bred from; accidents, however, are not to be regarded as unsoundness; but if a defect exists in a mare which is intended for breeding, the breeder should be certain that such defect is a mere accident, and not a natural malformation. Both parents should also be as free from moral as from physical infirmity; from faults or vices of temper or disposition. Although a defect of one parent may sometimes be counteracted by a preponderating excellence relative to that defect in the other, great care is necessary that both parents do not possess the same defect. If one would be perfectly certain in breeding, it is better to avoid even such mares as have suffered merely from accident; nor should the mare be put to the horse at too early an age, if one would avoid the hazard of obtaining an unreliable offspring.

The best form of a mare from which to breed, for any purpose, is a short-legged, lengthy animal, with a deep, roomy chest and carcass, wide and capacious hips, and a sound constitution. "Breed," of course, must be looked for, according to the class of horses to which the mare belongs; a good, animated countenance, an upright, sprightly carriage; general structure of muscle, bone and sinew firm, dense, and compact. The head of the brood-mare is an important point to be regarded; a mare that has a heavy head and a stupid countenance cannot breed a good foal, unless to a horse possessed of fire almost to madness—for her countenance indicates her disposition. The neck should be brought out of the top of the withers, and not of the bottom of the shoulders and chest; the shoulders should be well back, the blade-bone lying obliquely from the shoulder joint; the blade should also be long and wide, extending nearly to the top of the withers, but attached so closely and so well covered with muscle as not to present any undue prominence; the back of the shoulder should also be well furnished with muscle, appearing to the mounted rider of a wedge shape widening towards his knee; the fore-leg should be perpendicular, the toe and the point of the shoulder being in a right line; the foot should be round, even, and of a dark color; the heels should be open, but not low; the brisket should be deep, especially in the case of a riding-horse, as otherwise a crupper will be required to keep the saddle in its proper place; the quarters should be long and oval on the top; the hips cannot be too broad in a brood-mare, though in a stallion too wide hips are objectionable; the hocks should be regarded, and the shank-bone and sinew, both before and behind, should be well developed, and dropped straight below the joint.

As to the shape of the stallion little satisfactory can be said. It must depend upon that of the mare, and the kind of horse wished to be bred; but, if there is one point absolutely essential, it is compactness—as much goodness and strength as possible condensed into a small space. Next to compactness, the inclination of the shoulder should be regarded. A huge stallion, with upright shoulders, never got a capital hunter or hackney; from such nothing but a cart or dray horse can be obtained, and that, perhaps, spoiled by the opposite form of the mare. If, however, a merely slow draught-horse is desired, an upright shoulder is desirable, if not absolutely necessary.

THE ARAB STALLION JUPITER.

The principal requirements in connection with breeding may be concisely summed up as follows:—

First. There should be mutual adaptation in form and size, and indeed in all important characteristics, between the sire and the dam.

Second. If the mare be defective in any particular, she should not be bred to a stallion having a similar, or even an opposite, fault; but one should rather be chosen perfect in that point.

Third. Exceedingly small mares should not be bred with enormously large horses; distortions will generally be the result. For a mare of sixteen hands, a horse of not less than fifteen hands should be selected; if she be too low or small, the horse may be an inch or two higher, but not of the tall or leggy kind.

Fourth. As it is frequently the case, that without any known cause the blood of a certain kind of horses will not cross well with that of another, such instances when ascertained should be avoided.

Fifth. If the mare is of a good kind of horses, but one which has degenerated in size from "in-breeding," (that is, from continuous breeding into the same family and blood—with their own daughters and grand-daughters, in other words—for about two generations,) the only remedy is, to breed to the purest stallion that can be found, but of a different kind from hers, unless some ten or more generations removed.

Sixth. After breeding for several generations from males and females of one kind, it is generally beneficial to change to another entirely different; otherwise degeneracy in size will be the general result.

The mare should not be put to horse under three years of age. Although some contend that, if lightly worked, she may be used for breeding until she is twenty, yet it is very doubtful whether breeding from any mare over twelve years old, at the very utmost, will prove satisfactory. If a large colt is desired, have a large mare; as her size has generally more to do with the matter than that of the stallion. The most favorable time for putting the mare to the horse is from March to the beginning of May; colts foaled in March are generally found to turn out hardier, and to stand better, than those foaled earlier.

From the time of covering to within a few days of the expected period of foaling, the cart-mare may be kept at moderate labor not only without injury, but with decided advantage. She should then be released from work and kept near home under the frequent inspection of some careful person. When nearly half the time of pregnancy has elapsed, she should have a little better food, being allowed one or two feeds of grain in the day. As this is about the time when they are accustomed to slink their foals, or when abortion occurs, the owner's eye should be frequently upon her. Good feeding and moderate exercise are the best preventives of this mischance. As the mare that has once slunk her foal is liable to a repetition of this accident, she should never be suffered to be with other mares between the fourth and fifth months; for so great is the power of sympathy or imagination in the mare that if one suffers abortion, others in the same pasture will too often share the same fate. Farmers frequently suppose that such mishaps originate from some infection; and many wash and paint and tar their stables to prevent an infection that really lies in the imagination.

The period of pregnancy varies from forty-four to fifty-six weeks, but it is usually from forty-seven to fifty. If the mare, whether of pure or common breed, be cared for as suggested above, and be in good health while in foal, little danger will attend the act of giving birth to the young. Should there be, however, false presentation of the fœtus, or any difficulty in producing it, recourse should be had to a well-informed veterinary surgeon, rather than to run the risk of injuring the mare by violent attempts to relieve her.

After the mare has foaled, she should be turned into some well-sheltered pasture, with a shed or hovel into which she may run when she pleases. If she has foaled early and the grass is scanty, she should have a feed or two of oats or Indian corn daily; if the corn is given in a trough upon the ground, the foal will partake of it with her. Nothing is gained at this time by starving the mare and stinting the foal. When the new grass is plentiful, the quantity of grain may be gradually diminished. The proper care of young foals will repay a hundred-fold; this being, indeed, the most critical period of the animal's life, when attention or neglect produces the most noticeable and permanent results.

If convenient, the foal may be permitted to run for twelve months at the foot of the mare; but when mares are kept expressly for breeding purposes, many circumstances render this objectionable. Within about a month or six weeks from foaling the mare will be again in heat, and should be put to the horse; at the same time, also, if she is used for agricultural purposes, she may resume light work. At first, the foal should be shut up in the stable during working hours; but, as it acquires sufficient strength, it is better to allow it to follow its dam. The work will contribute to the health of the mother, and increase her flow of milk; and the foal, by accompanying her, will suck more frequently, thrive better, become tamed, and gradually familiarized with the objects among which it is afterward to live. While the mare is thus worked, she and the foal should be well fed; and two feeds of corn, at least, should be added to the green food which they get when turned out after their work, and at night.

In five or six months, according to the growth of the foal, it may be weaned. For this purpose, it should either be housed, or turned into some pasture at a distance from the dam. The mare should be put to harder work and drier food. If her milk is troublesome, or she pines after her foal, a few purgatives (one or two urine-balls, or a physic ball) will be found useful. The foal should be fed well and liberally every morning and evening, bruised oats and bran being about the best kind of food which can be given. The money so laid out upon the liberal nourishment of the colt, is well expended; yet, while he is well fed, he should not be rendered delicate by excess of care. Toward the end of summer the foal may be turned out to general pasture without fear of his again seeking his dam.

Should the foal be a male, and emasculation be desirable, it is better to perform the operation at the time of weaning, that the one trouble shall serve for both occasions. If, however, weaning take place in June or July, when the fly abounds, the operation should not be performed, as this insect by its attacks will cause restlessness and consequent inflammation, and thus retard recovery. Early spring, or an advanced period of autumn, is the best time. This operation should in no instance be performed by any other than a competent veterinary surgeon. One thing in this connection should be mentioned; when a horse is suffered to attain two-thirds of his growth before emasculation, an animal is obtained of form, power, and value far superior to that which has been operated upon when a foal. This much is deserving of remembrance; though we cannot omit heartily condemning the practice of emasculation at all.

BREAKING.

No greater mistake can be made than the postponement of this part of the rearing of a horse. It should always commence as soon as the colt is weaned, or immediately after the effects of the emasculation have disappeared; it should in this manner be commenced and carried on gradually, with gentleness and kindness. The foal should be daily handled, partially dressed, accustomed to the halter when led about, and even tied up occasionally for an hour or so.

BREAKING.

The tractability, good temper, and value of the horse depend much more upon this than most breeders consider. The person who feeds the colt should have the entire management of him at this period, and he should be a trustworthy person, possessed of a quiet, uniform temper and a kindly disposition. Many a horse is spoiled and rendered permanently untamable by early harshness or improper treatment; and many a horse that otherwise would have proved a vicious, unmanageable brute, has been brought to be a docile, gentle, and affectionate servant by the judicious treatment of those to whose charge his management at this particular period was fortunately intrusted.

Such a treatment is sufficient for the first year; after the second winter, the operation of training should commence in good earnest. The colt should be bitted, a bit being selected which will not hurt his mouth, and much smaller than those in common use. The work of bitting may perhaps occupy three or four days; the colt being suffered to amuse himself with the bit, to play, and to champ it for an hour or so during a few successive days. When he has become accustomed to the bit, he may have two long ropes attached to it, slightly fastened to his sides by a loose girth over the back, and his feeder may thus drive him, as it were, around a field, pulling upon him as he proceeds. This will serve as a first lesson in drawing. If he is intended for a saddle-horse, a filled bag may be thrown across his back and there secured, and, after he has become used to this, a crotch may be fastened upon his back, its lower extremities grasping his sides, and thus preparing him for the legs of his rider.

Portions of the harness may next be put upon him, reserving the blind winkers for the last; and a few days afterward he may go into the team. It is better that he should be one of three horses, having one before him, and the shaft-horse behind him. There should at first be the mere empty wagon; and the draught is best begun over the grass, where the colt will not be frightened by the noise of the wheels. Nothing should be done to him, except giving him an occasional pat or a kind word. The other horses will keep him moving and in his place; and after a short time, sometimes even during the first day, he will begin to pull with the rest. The load may then be gradually increased.

If the horse is desired for purposes of riding as well as for exclusively agricultural uses, his first lesson may be given when he is in the team; his feeder, if possible, being the first one put upon him. He will be too much confined by the harness and by the other horses, to make much resistance; and, in the greater number of instances, will quietly and at once submit Every thing, however, should proceed gradually and by successive steps, and, above all, no whip or harsh language should, under any circumstances, be allowed to be used. Although mildness is absolutely essential, it is none the less necessary that the colt should be taught implicit obedience to the will of his master. To accomplish this, neither whip, nor spur, nor loud shouting, nor hallooing is necessary; the successful horsebreaker is required to possess but the three grand requisites of firmness, steadiness, and patience.

When the colt begins to understand his business somewhat, the most difficult part of his work, backing, may be taught him; first, to back well without anything behind him, then with a light curb, and afterwards with some more heavy load—the greatest possible care being always taken that his mouth be not seriously hurt. If the first lesson causes much soreness of the gums, he will not readily submit to the second. If he has been previously rendered tractable by kind usage, time and patience will accomplish every thing that is desired. Some persons are in the habit of blinding the colt when teaching him to back. This can only be necessary with a restive and obstinate one, and even then should be used only as a last resort.

In the whole process of breaking it should constantly be borne in mind, that scarcely any horses are naturally vicious. Cruel usage alone first provokes resistance. If that resistance is followed by greater severity, the stubbornness of the colt increases in proportion; open warfare ensues, in which the man seldom gains the advantage, and the horse is frequently rendered utterly unfit for service. Correction may, indeed, be necessary for the purpose of enforcing implicit obedience, after the training has proceeded to a certain extent; but the early lessons should be imparted with kindness alone. Young colts are sometimes very perverse; and many days will occasionally pass, before they will suffer the bridle to be put on, or the saddle to be worn. It must not, however, be forgotten, that a single act of harshness will indefinitely increase this length of time; but that patience and kindness will always prevail. On some occasion, when the colt is in a better humor than usual, the bridle may be put on, or the saddle be worn; and, if this compliance, on his part is accompanied by kindness and soothing on the part of the breaker, and no inconvenience or pain be suffered by the animal, all resistance will be ended.

The same principles will apply to the breaking-in of the horse for the road. The handling and some portion of instruction should commence from the time of weaning; for upon this the future tractibility of the horse in a great measure depends. At two years and a half, or three years, the regular process of breaking-in should commence. If it is put off until the animal is four years old, his strength and obstinacy will be more difficult to overcome. The plan usually adopted by the breaker cannot, perhaps, be much improved; except that there should be much more kindness and patience, and far less harshness and cruelty, than those persons are accustomed to exhibit, and a great deal more attention to the form and natural action of the horse. A headstall is put on the colt, and a cavesson (or apparatus to confine and pinch the nose,) affixed to it with long reins. He is first accustomed to the rein, then led around a ring on soft ground, and at length mounted and taught his paces. Next to preserving the temper and docility of the horse, there is nothing of so much importance, as to teach him every pace and every part of his duty distinctly and thoroughly. Each should constitute a separate and sometimes long-continued lesson, taught by a man who will never allow his passion to overmaster his discretion.

After the cavesson has been attached to the headstall, and the long reins put on, the colt should be quietly led about by the breaker—a steady boy following behind, to keep him moving by occasional threatening with the whip, but never by an actual blow. When the animal follows readily and quietly, he may be taken to the ring and walked around, right and left, in a very small circle. Care should be taken to teach him this pace thoroughly, never allowing him to break into a trot. The boy with his whip may here again be necessary, but an actual blow should never be inflicted.

Becoming tolerably perfect in the walk, he should be quickened to a trot, and kept steadily at it; the whip and the boy, if needful, urging him on, and the cavesson restraining him. These lessons should be short, the pace being kept perfect and distinct in each, and docility and improvement rewarded with frequent caresses, and handfuls of corn. The length of the rein may now be gradually increased, and the pace quickened, and the time extended, until the animal becomes tractable in these his first lessons; toward the conclusion of which, crupper straps, or something similar, may be attached to the clothing. These, playing about the sides and flanks, accustom him to the flapping of the coat of the rider. The annoyance which they occasion will pass over in a day or two; for when the animal learns by experience that no harm comes from them, he will cease to regard them.

Next comes the bitting. The bits should be large and smooth, and the reins buckled to a ring on each side of the pad. There are many curious and expensive machines for this purpose, but the simple rein will be quite sufficient. It should at first be slack, and then very gradually tightened. This prepares for the more perfect manner in which the head will afterward be got in its proper position, when the colt is accustomed to the saddle. Occasionally the breaker should stand in front of the colt, and take hold of each side-rein near to the mouth, and press upon it, and thus begin to teach him to stop and to back on the pressure of the rein, rewarding every act of docility, and not being too eager to punish occasional carelessness or waywardness.

The colt may now be taken into the road or street, that he may become gradually accustomed to the objects among which his services will be required. Here, from fear or playfulness, a considerable degree of starting and shying may be exhibited, of which as little notice as possible should be taken. The same or a similar object should be soon passed again, but at a greater distance. If the colt still shies, let the distance be still further increased, until he takes no notice of the object. Then he may be gradually brought nearer to it; and this may usually be accomplished without the slightest difficulty; whereas, had there been an attempt to force him close to it in the first instance, the remembrance of the contest would have been associated with every appearance of the object, and the habit of shying would have been established.

Hitherto, with a cool and patient breaker, the whip may have been shown, but will scarcely have been used; the colt should now, however, be accustomed to this necessary instrument of authority. Let the breaker walk by the side of the animal, throw his right arm over his back, holding the reins in his left, occasionally quickening his pace, and at the moment of doing this tapping the horse with the whip in his right hand, and at first very gently. The tap of the whip and the quickening of the pace will soon become associated in the animal's mind. If necessary, these reminders may gradually fall a little heavier, and the feeling of pain be the monitor of the necessity of increased exertion. The lessons of reining in and stopping, and backing on the pressure of the bit, may continue to be practised at the same time.

He may next be taught to bear the saddle. Some little caution will be necessary in first putting it on. The breaker should stand at the head of the colt, patting him and engaging his attention, while one assistant, on the off-side, gently places the saddle on the back of the animal; another on the nearest side slowly tightening the girths. If he submits quietly to this, as he generally will when the previous process of breaking-in has been properly conducted, the operation of mounting may be attempted on the following, or on the third day. The breaker will need two assistants in order to accomplish this. He will remain at the head of the colt, patting and making much of him. The rider will put his foot into the stirrup, and bear a little weight upon it, while the man on the opposite side presses equally on the other stirrup-leather; and according to the docility of the animal, he will gradually increase the weight, until he balances himself on the stirrup. If the colt is uneasy or fretful, he should be spoken kindly to and patted, or a mouthful of grain be given to him; but if he offers serious resistance, the lessons must terminate for that day. He may possibly be in a better humor on the morrow.

When the rider has balanced himself for a minute or two, he may gently throw his leg over, and quickly seat himself in the saddle. The breaker should then lead the animal around the ring, the rider meanwhile sitting perfectly still. After a few minutes he should take the reins, and handle them as gently as possible, guiding the horse by the pressure of them; patting him frequently, and especially when he thinks of dismounting; and, after having dismounted, offering him a little grain, or green feed. The use of the rein in checking him, and of the pressure of the leg and the touch of the heel in quickening his pace, will soon be taught, and his education will be nearly completed.

The horse having thus far submitted himself to the breaker, these pattings and awards must be gradually diminished, and implicit obedience mildly but firmly enforced. Severity will not often be necessary, in the great majority of cases it being altogether uncalled for; but should the animal, in a moment of waywardness, dispute the command of the breaker, he must at once be taught that he is the slave of man, and that we have the power, by other means than those of kindness, to bend him to our will. The education of the horse, in short, should be that of the child. Pleasure is, as much as possible,

THE AGRICULTURIST'S METHOD.

associated with the early lessons; but firmness, or, if need be, coercion, must establish the habit of obedience. Tyranny and cruelty will, more speedily even in the horse than in the child, provoke the wish to disobey; and, on every practicable occasion, the resistance to command. The restive and vicious horse is, in ninety-nine cases out of a hundred, made so by ill-usage, and not by nature. None but those who will take the trouble to make the experiment, are aware how absolute a command the due admixture of firmness and kindness will soon give us over any horse.

CASTRATION.

The period at which this operation may be best performed depends, as has been previously remarked, much on the breed and form of the colt, and the purpose for which he is destined. For the common agricultural horse, the age of four or five months will be the most proper time, or, at least before he is weaned. Few horses are lost when cut at that age; though care should be taken that the weather is not too bad, nor the flies too numerous.

If the horse is designed either for the carriage or for heavy draught, he should not be castrated until he is at least a year old; and, even then, the colt should be carefully examined. If he is thin and spare about the neck and shoulders, and low in the withers, he will materially improve by remaining uncut another six months; but if his fore quarters are fairly developed at twelve months, the operation should not be delayed, lest he grow gross and heavy before, and, perhaps, has begun too decidedly to have a will of his own. No specific age, therefore, can be fixed; but the operation should be performed rather late in the spring, or early in the autumn, when the air is temperate and particularly when the weather is dry.

No preparation is necessary for the sucking colt, but it may be prudent to physic one of more advanced age. In the majority of cases, no after treatment will be necessary, except that the animal should be sheltered from intense heat, and more particularly from the wet. In temperate weather he will do much better running in the field than nursed in a close and hot stable. The moderate exercise which he will necessarily take in grazing, will be preferable to entire inaction.

The old method of opening the *scrotum*, or testicle bag, on each side, and cutting off the testicles, preventing bleeding by a temporary compression of the vessel, while they are seared with a hot iron, must not, perhaps, be abandoned; but there is no necessity for that extra pain, when the spermatic cord (the blood-vessels and the nerve,) is compressed between two pieces of wood as tightly as in a vice, and there left until the following day, when it may be removed with a knife.

The practice of some farmers of cording, or twitching their colts at an early period exposes the animal to much unnecessary pain, and is attended with no slight danger.

Another method of castration is by *torsion*. An incision is made into the *scrotum*, and the *vas deferens* is exposed and divided. The artery is then seized by a pair of forceps contrived for the purpose, and twisted six or seven times round. It retracts without untwisting the coils, and bleeding ceases. The testicle is removed, and there is no sloughing or danger. The most painful part of the operation—the operation of the firing-iron, or the claws—is avoided, and the wound readily heals. It is to be remarked, in this connection, that the use of chloroform has been found very beneficial in performing the operation in the old way, both in removing all pain, and also preventing that severe struggling which often takes place, and which has sometimes been followed with very dangerous consequences. With the assistance of this agent, the operation has been safely performed in seven minutes, without any pain to the animal.

DOCKING.

This is an operation, whose only sanction is to be found in the requirements of a senseless fashion. "The convenience of the rider," which is sometimes urged in its favor, is the veriest nonsense afloat. In truth, the operation is one of the most useless which the brain of man, fertile in romance and expedients as it is, ever devised; since, instead of adding to the beauty of the animal, as some assert, it but adds deformity. Not many years back, this attempted improvement upon nature became a perfect mania. In England, however, this cruel practice has been almost entirely discarded; and it is to be hoped that the operation in the United States also will speedily be frowned down.

The operation, as now performed by veterinary surgeons, was introduced some years ago by the American Veterinary Association of Philadelphia. It consists in passing a narrow-bladed knife (a pricking knife will answer,) between the coccygeal bones at the desired point, from above downwards, cutting outwards and backwards on each side so as to form two flaps, which are carefully brought together over the end of the tail and secured by the interrupted suture; thus giving protection to the stump of the tail, and making a much neater finish than by any other method which could be adopted. No styptic whatever is required, and there need be no fear of hemorrhage, as the union generally takes place by what surgeons call first intention. If, however, the flaps do not fit nicely, healing will not take place without suppuration. This fact should be borne in mind in performing the operation, as much time in healing may thus be saved.

By the old method that joint is searched for, which is nearest to the desired length of tail. The hair is then turned up, and tied round with tape for an inch or two above this joint, and that lying immediately upon the joint is cut off. The horse is fettered with the side-line, and then the veterinary surgeon with his docking-machine, or the farmer with his carving-knife and mallet, cuts through the tail at one stroke.

Some farmers dock their colts a few days after they are dropped. This is a commendable custom on the score of humanity. No colt was ever lost by it; the growth of the hair, and the beauty of the tail not being at all impaired.

NICKING.

This barbarous operation was once sanctioned by fashion, and the breeder and the dealer are even now sometimes tempted to inflict the torture of it in order to obtain a ready sale for their colts. It is not, practiced to the extent that it used to be, nor is it attended by so many circumstances of cruelty.

The operation is thus performed. The side-line is put on the horse, or some persons deem it more prudent to cast him, and that precaution may be recommended. The hair at the end of the tail is securely tied together, for the purpose of afterward attaching a weight to it. The operator then grasps the tail in his hand, and, lifting it up, feels for the *centre* of one of the bones—the prominences at the extremities guiding him—from two to four inches from the root of the tail, according to the size of the horse. He then with a sharp knife divides the muscles deeply from the edge of the tail on one side to the centre, and, continuing the incision across the bone of the tail, he makes it as deep on the other side. One continued incision, steadily yet rapidly made, will accomplish all this. If it is a blood-horse that is operated on, this will be sufficient. For a hunter, two incisions are usually made, the second being about two inches below the first, and likewise as nearly as possible in the centre of one of the bones.

On a hackney, a third incision is made; for fashion has decided that his tail shall be still more elevated and curved. Two incisions only are made in the tail of a mare, and the second not very deep.

When the second incision is made, some fibres of the muscles between the first and second will project into the wound, and must be removed by a pair of curved scissors. The same must be done with the projecting portions from between the second and third incisions. The wound should then be carefully examined, in order to ascertain that the muscles have been equally divided on each side, otherwise the tail will be carried awry. This being done, pieces of tow must be introduced deeply into each incision, and confined, but not too tightly, by a bandage. A very profuse bleeding only will justify any tightness of bandage, and the ill consequences that have resulted from nicking are mainly attributable to the unnecessary force that is used in confining these pledgets of tow. Even if the bleeding, immediately after the operation, should have been very great, the roller must be loosened in two or three hours, otherwise swelling and inflammation, and even death, may possibly ensue. Twenty-four hours after the operation, the bandage must be quite removed; and then all that is necessary, so far as the healing of the incisions is concerned, is to keep them clean.

The wounds must remain open; and this can only be accomplished by forcibly keeping the tail curved back during two or three weeks. For this purpose, a cord, one or two feet in length, is affixed to the end of the hair, which terminates in an-

THE USUAL METHOD.

other divided cord, each division going over a pulley on each side of the back of the stall. A weight is hung at each extremity, sufficient to keep the incisions properly open, and regulated by the degree in which this is wished to be accomplished. The animal will thus be retained in an uneasy position, although, after the first two or three days, probably not of acute pain. It is barbarous to increase this uneasiness or pain by affixing too great a weight to the cords; for it should be remembered that the proper elevated curve is given to the tail, not by the weight's keeping it in a certain position for a considerable time, but by the depth of the first incisions, and the degree in which the wounds are kept open.

The dock should not, for the first three or four days, be brought higher than the back. Dangerous irritation and inflammation would probably otherwise be produced. It may, after that, be gradually raised to an elevation of forty-five degrees. The horse should be taken out of the pulleys, and gently exercised once or twice every day; but the pulleys cannot finally be dispensed with until a fortnight after the wounds have healed, because the process of contraction, or the approach of the divided parts, goes on for some time after the skin is perfect over the incisions, and the tail would thus sink below the desired elevation. The French method is simpler and less barbarous than ours, allowing the horse to lie down or move about at his pleasure. Where this operation is to be performed, it might be adopted with advantage as shown in the engraving annexed.

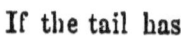

THE FRENCH METHOD.

If the tail has not been unnecessarily extended by enormous weights, no bad consequences will usually follow; but if considerable inflammation should ensue, the tail must be taken from the pulley, and carefully fomented with simple warm water, and a dose of physic given Locked-jaw has, in some rare instances, followed, under which the horse generally perishes. The best means of cure in the early state of this disease, is to amputate the tail at the joint above the highest incision. In order to prevent the hair from coming off, it should be unplaited and combed out every fourth or fifth day.

THE STABLE.

The most desirable thing in a stable is *ventilation*. A horse requires air equally with his master; and as the latter requires a chimney to his sleeping apartment, so does the former. This may be a mere outlet through the ceiling, or it may be formed as a dome or cupola. It should not, of course, be open at the top, or rain will get in, but roofed over, and have an opening at the sides. Besides this, there should be openings in the wall, near the ground, but not in the stalls. This will produce a thorough air, and may be so placed as not to expose the horses to the draught.

The stable should not be less than twelve feet high, from floor to ceiling, and the floor should be well paved, slope slightly backward, and along the back of the stalls should run a gutter, about a foot wide and an inch or two deep. No stable should be less than eighteen feet deep, and each stall should be at least six feet clear; but if eight feet can be allowed, so much the better. Although some horses will agree when kept together in one stall, it is far preferable to allow each a stall to himself. The manger should be about sixteen inches deep, the same from front to back, more narrow at bottom than at top, and two feet in length. The rack is best when closed in front, the back part being an inclined plane of wood sloping gradually toward the front, and terminating about two feet down. This kind of rack effects a considerable saving in hay; for the reader scarcely needs to be reminded that in the common rack much of the hay given is dragged down and trampled in the litter. It also prevents the hay-seed from falling into the horse's eyes; for the rack is on a level with the manger, and about three feet from the ground. Another advantage gained by this rack is the facility with which it can be filled, thus obviating all necessity for a loft over the stable, and, consequently, admitting of a greater height of ceiling above the horses, as well as of a superior ventilation.

CUSTOMARY FORM OF STALLS.

The windows and the doors should be at opposite ends, as ventilation is thereby promoted; the doors should be divided transversely, at the height of about four feet from the ground. The upper portion may thus be occasionally opened. Whitewash is a bad dressing for the interior of the stable, as it causes too great a glare of light; paint of a leaden color is best, and it can be washed from time to time with soap and water. There should be a bin, properly divided into partitions for oats, beans, and the like; and this is better at the back of the stable.

A few buckets of water dashed over the floor of the stable while the horses are at work, will keep all sweet. The litter should also be turned out to dry, and a little fresh straw spread for the horses to stale on. A shed placed beside the stable is a great advantage, on two accounts—it admits of the litter being dried, and the horse dressed there in wet and stormy weather.

A little powdered gypsum, strown upon the stable floor, will also act by absorbing the ammoniacal gas, and thus removing its foul smell—a frequent predisposing cause of ophthalmia. If the ammonia, however, accumulates in any considerable quantity, the speediest and most efficacious remedy as a disinfectant is muriatic acid.

AIR.

The importance of thorough ventilation has been adverted to under the preceding head, but a few words additional seem necessary.

A hot stable has in the minds of many been long connected with a glossy coat for the horse. The latter, it is thought, cannot be obtained without the former. To this it may be replied that in winter a thin, glossy coat is not desirable. Nature gives to every animal a warmer clothing when the cold weather approaches. The horse—the agricultural horse, especially—acquires a thicker and a lengthened coat, in order to defend him from the surrounding cold. Man puts on additional and a warmer covering, and his comfort is increased and his health preserved by it. He who knows anything of the farmer's horse, or cares about his enjoyment, will not object to a coat a little longer, and a little roughened when the wintry wind blows bleak. The coat, however, does not need to be so long as to be unsightly; and warm clothing, even in a cool stable, will, with plenty of careful and faithful grooming, keep the hair sufficiently smooth and glossy to satisfy the most fastidious. The over-heated air of a close stable dispenses with the necessity of this grooming, and therefore the idle attendant unscrupulously sacrifices the health and safety of the horse.

If the stable is close, the air will not only be hot, but foul. The breathing of every animal contaminates it; and when, in the course of the night, with every aperture stopped, it passes again and again through the lungs, the blood cannot undergo its proper and healthy change; digestion will not be so perfectly performed, and all the functions of life are injured. Let the owner of a valuable horse think of his passing twenty or twenty-two out of the twenty-four hours in this debilitating atmosphere. Nature does wonders in enabling every animal to accommodate itself to the situation in which it is placed, and the horse that lives in the stable-oven suffers less from it than would scarcely be deemed possible; but he does not, and cannot, possess the power and hardihood which he would acquire under other circumstances.

The air of the improperly closed and heated stable is still further contaminated by the urine and dung, which rapidly ferment there, and give out stimulating and unwholesome odors. When one first enters an ill-managed stable, and especially early in the morning, he is annoyed, not only by the heat of the confined air, but by a pungent smell, resembling hartshorn; what surprise, then, need be excited at the inflammation of the eyes, and the chronic cough, and the disease of the lungs, by which the animal, which has been all night shut up in this vitiated atmosphere, is often attacked; or if glanders and farcy should occasionally break out in such stables? Chemical experiments have demonstrated that the urine of the horse contains in it an exceedingly large quantity of hartshorn; and not only so, but that, influenced by the heat of a crowded stable, and possibly by other decompositions which are going on at the same time, this ammoniacal vapor begins to be rapidly given out almost immediately after the urine is voided.

When disease begins to appear among the inhabitants of these ill-ventilated places, it is not wonderful that it should rapidly spread among them, and that the plague-spot should be, as it were, placed on the door of such a stable. When distemper appears in spring or autumn, it is in very many cases to be traced to such a pest-house. It is peculiarly fatal there. The horses belonging to a small establishment, and rationally treated, have it comparatively seldom, or, when they do, but lightly; but among the inmates of a crowded stable it is sure to display itself, and there it is most deadly. The experience of every veterinary surgeon, and of every large proprietor of horses, will corroborate this statement.

Every stable, then, should possess within itself a certain degree of ventilation. The cost of this would be trifling, and its saving in the preservation of valuable animals may be immense. The apertures need not be large, and the whole, as before said, may be so contrived that no direct current of air shall fall on the horse.

A gentleman's stable should never be without a thermometer. The temperature should seldom exceed seventy degrees in the summer, or sink below forty or fifty degrees in the winter.

LITTER.

Having spoken of the vapor of hartshorn, which is so rapidly and plentifully given out from the urine of a horse in a heated stable, the subject of litter comes naturally next in order. The first caution is, frequently to remove it. The early extrication of gas shows the rapid putrefaction of the urine; the consequence of which will be the rapid putrefaction of the litter which is moistened by it. Every thing hastening to decomposition should be carefully removed where life and health are to be preserved. The litter which has been much wet or at all softened by the urine, and is beginning to decay, should be swept away every morning; the greater part of the remainder may then be piled under the manger; a little being left to prevent the painful and injurious pressure of the feet on the hard pavement during the day. The soiled and soaked portion of that which was left should be removed at night. In the better kind of stables, however, the stalls should be completely emptied every morning.

No heap of fermenting dung should be suffered to remain during the day in the corner or in any part of the stable. With regard to this, the directions for removal should be peremptory.

The stable should be so contrived that the urine may quickly run off, and the offensive and injurious vapor from the decomposing fluid, and the litter will thus be materially lessened; but if this is effected by means of gutters and a descending floor, the descent must be barely sufficient to cause the fluid to escape, as, if the toes are kept higher than the heels, it will lead to lameness, and is also a frequent cause of contraction of the foot.

Stalls of this kind certainly do best for mares; but for horses those are preferable, which have a grating in the centre, and a slight inclination in the floor on every side towards the middle. A short branch may communicate with a larger drain, by means of which the urine may be carried off to a reservoir outside the stable. Traps are now contrived, and may be procured at little expense, by means of which neither any offensive smell nor current of air can pass through the grating.

In stables with paved floors particularly, humanity and interest, as well as the appearance of the stable, should induce the proprietor of the horse to place a moderate quantity of litter under him during the day.

LIGHT.

This neglected branch of stable-management is of far more consequence than is generally imagined. The farmer's stable is frequently destitute of any glazed window, and has only a shutter, which is raised in warm weather, and closed when the weather becomes cold. When the horse is in the stable only during a few hours in the day, this is not of so much consequence, nor of so much, probably, with regard to horses of slow work; but to carriage-horses and roadsters, so far, at least, as the eyes are concerned, a dark stable is little less injurious than a foul and heated one. In order to illustrate this, reference may be made to the unpleasant feeling, and the utter impossibility of seeing distinctly, when a man suddenly emerges from a dark place into the full glare of day. The sensation of mingled pain and giddiness is not speedily forgotten; and some minutes elapse before the eye can accustom itself to the increased light. If this were to happen every day, or several times in a day, the sight would be irreparably injured, or possibly blindness would be the final result. We need not wonder, then, that the horse, taken from a dark stable into a blaze of light, feeling, probably, as we should do under similar circumstances, and unable for any time to see anything around him distinctly, should become a starter; or that the frequently repeated violent effect of sudden light should induce inflammation of the eye so intense as to terminate in blindness. There is, indeed, no doubt that horses kept in dark stables are frequently notorious starters, and that abominable habit has been properly traced to this cause.

If plenty of light be admitted, the walls of the stable, and especially that portion of them which is before the horse's head, must not be of too glaring a color. The color of the stable should depend on the quantity of light. Where much can be admitted, the walls should be of a gray hue. Where darkness would otherwise prevail, frequent painting may in some degree dissipate the gloom.

For another reason, it will be evident that the stable should not possess too glaring a light; it is the resting-place of the horse. The work of the farmer's horse, indeed, is principally confined to the day. The hours of exertion having passed, the animal returns to his stable to feed and to repose, and the latter is as necessary as the former, in order to prepare him for renewed work. Something like the dimness of twilight is requisite to induce the animal to compose himself to sleep. This half-light is more particularly adapted to horses of heavy work. In the quietness of a dimly-lighted stable, they obtain repose, and accumulate flesh and fat.

GROOMING.

To the agriculturist it is not necessary to say much under this head, as custom, apparently without any ill effect, has allotted so little of the comb and brush for the farmer's horse. The animal that is worked all day, and turned out at night,

GROOMING.

requires little more to be done to him than to have the dirt brushed off his limbs. Regular grooming, by rendering his skin more sensitive to the alteration of temperature, and the inclemency of weather, would be prejudicial. The horse that is altogether turned out, needs no grooming. The dandruff, or scurf, which accumulates at the roots of the hair, is a provision of nature to defend him from the wind and the cold.

It is to the stabled horse, highly fed, and little or irregularly worked, that grooming is of so much consequence. Good rubbing with the brush, or the curry-comb, opens the pores of the skin, causes the blood to circulate to the extremities of the body, produces free and healthy perspiration, and stands in the stead of exercise. No horse will carry a fine coat without either unnatural heat, or dressing. They both effect the same purpose; they both increase the insensible perspiration; but the first does it at the expense of health and strength, while the second, at the same time that it produces a glow on the skin, and a determination of blood to it, rouses all the energies of the frame. It would be well if the proprietor of the horse were to insist—and to see that his orders are implicitly obeyed —that the fine coat, in which he and his groom so much delight, is produced by honest rubbing, and not by a heated stable and thick clothing, and, most of all, not by stimulating or injurious spices. The horse should be regularly dressed every day, in addition to the grooming that is necessary after work.

When the weather will permit the horse to be taken out, he should never be groomed in the stable, unless he is an animal of peculiar value, or placed for a time under peculiar circumstances. Without dwelling on the want of cleanliness, when the scurf and dust which are brushed from the horse lodge in his manger, experience teaches that, if the cold is not too great, the animal is braced and invigorated to a degree that cannot be attained in the stable, from being dressed in the open air. There is no necessity, however, for half the punishment which is often inflicted upon the horse in the act of dressing; and particularly on one whose skin is thin and sensitive. The curry-comb should always be applied lightly. With many horses, its use may be almost dispensed with; and even the brush does not need to be so hard, nor the points of the bristles so irregular as they often are. A soft brush, with a little more weight of the hand, will be equally effectual, and much more pleasant to the horse. A hair-cloth, while it will seldom irritate and tease, will be almost sufficient with horses that have a thin skin, and that have not been neglected. After all, it is no slight task to dress a horse as it ought to be done. It occupies no little time, and demands considerable patience, as well as dexterity. It will be readily ascertained whether a horse has been well dressed, by rubbing him with one of the fingers. A greasy stain will expose the idleness of the groom. When, however, the horse is changing his coat, both the curry-comb and the brush should be used as lightly as possible.

Whoever would be convinced of the benefit of friction to the horse's skin, and to the horse generally, needs only to observe the effects produced by rubbing the legs of a tired horse well with the hands. While every enlargement subsides, and the painful stiffness disappears, and the legs attain their natural warmth and become fine, the animal is evidently and rapidly reviving; he takes hold of his food with zest, and then quietly lies down to rest.

EXERCISE.

The remarks upon this branch, also, can have but a slight reference to the agricultural horse. His work is usually, regular, and not exhausting. He is neither predisposed to disease by idleness, nor worn out by excessive exertion. He, like his master, has enough to do to keep him in health, and not enough to distress or injure him; on the contrary, the regularity of his work prolongs life to an extent seldom witnessed in the stable of the gentleman. These remarks on exercise, then, must have a general bearing, or have principal reference to those persons who keep a horse for business or pleasure, but cannot afford to maintain a servant for the express purpose of looking after it. The first rule to be laid down is, that every horse should have daily exercise. The animal, that with the usual stable feeding stands idle for three or four days, as is the case in many establishments, must suffer. He is predisposed to fever, or to grease, or, most of all, to diseases of the foot; and if, after three or four days of inactivity, he is ridden far and fast, he is almost sure to have inflammation of the lungs or of the feet.

Any horse, used for business or pleasure merely, suffers much more from idleness than he does from work. A stable-fed horse should have two hours' exercise every day, if he is to be kept free from disease. Nothing of extraordinary, or even of ordinary, labor can be effected on the road or in

EXERCISE.

the field, without sufficient and regular exercise. It is this which alone can give energy to the system, and develop the powers of any animal.

In training the race-horse, or the horse for hunting purposes, regular exercise is the most important of all considerations, however much it may be neglected in the usual management of the stable. The exercised horse will discharge his task, and sometimes a severe one, with ease and pleasure; while the idle and neglected one will be fatigued before half his labor is accomplished; and, if he is pushed a little too far, dangerous inflammation will ensue. How often, nevertheless, does it happen, that the horse which has stood inactive in the stable for three or four days, is ridden or driven thirty or forty miles in the course of a single day! The rest is often purposely given in order that he may be prepared for extra exertion—to lay in a stock of strength for the performance required of him—and then the owner is surprised and dissatisfied if the animal is fairly knocked up, or possibly becomes seriously ill. Nothing is so common and so preposterous, as for a person to buy a horse from a dealer's stable, where he has been idly fattened for sale for many a day, and immediately to give him a long run, and then to complain bitterly, and think that he has been imposed upon, if the animal is exhausted before the end, or is compelled to be led home suffering from violent inflammation. Regular and gradually increasing exercise would have made the same horse appear a treasure to his owner.

Exercise should be somewhat proportioned to the age of the horse. A young horse requires more than an old one. Nature has given to young animals of every kind a disposition to activity; but the exercise must not be violent. Much depends upon the manner in which it is given. To preserve the temper, and to promote health, it should be moderate, at least at the beginning and at the termination. The rapid trot, or even the gallop, may be resorted to in the middle of the exercise, but the horse should be brought in cool.

FOOD.

One half of the diseases of the horse owe their origin to over-feeding with hay. This applies more particularly to young horses, and to such as are not put to severe work. They are ever placed before a full rack, and, like children gorged with bread and butter, they eat merely for amusement, until at length the stomach gradually becomes preternaturally distended, the appetite increases in a relative proportion, becomes sooner or later voracious, and finally merges into a mere craving—it being a matter of indifference what the food is, so that the stomach is filled with it. This depravity of appetite is always accompanied by more or less thirst. This naturally enough produces general debility of the entire digestive function, including stomach, bowels, liver, spleen, and pancreas; worms are produced in thousands, and symptoms present themselves of so many varied hues, that enumeration, far less classification, becomes utterly impossible.

A horse's appetite is not to be taken as the criterion by which to determine the quantity of hay which he is to be permitted to consume; for most horses will eat three or four times as much as they ought. Horses have been known to consume thirty pounds weight of hay between a day and a night; and ten pounds is the most that should have been given during that time. Upon eight pounds of hay daily, with a due allowance of oats, a horse can be kept in full work, in prime health and spirits. It is better to keep young horses at grass until about five years old, and to work them during that period. When kept in the stable and not worked they are apt to acquire many very bad habits; and if the rack and manger be kept empty, with a view of preventing the over-loading of their stomachs, they will fall into a habit of playing with and mouthing them—a habit which finally degenerates into wind-sucking or crib-biting.

The system of manger-feeding is now becoming general among farmers. There are few horses that do not habitually waste a portion of their hay; and by some the greater part is pulled down and trampled under foot, in order first to cull the sweetest and best locks, which could not be done while the hay was confined in the rack. A good feeder will afterward pick up much of that which was thrown down: but some of it must be soiled and rendered disgusting, and, in many cases, one-third of this division of their food is wasted. Some of the oats and beans are imperfectly chewed by all horses, and scarcely at all by hungry and greedy ones. The appearance of the dung will sufficiently establish this.

The observation of this induced the adoption of manger-feeding, or of mixing a portion of cut feed with the grain and beans. By this means the animal is compelled to chew his food; he cannot, to any great degree, waste the straw or hay; the cut feed is too hard and too sharp to be swallowed without sufficient mastication, and while he is forced to grind that down, the oats and the beans are also ground with it, and thus yield more nourishment; the stomach is more slowly filled, and therefore acts better upon its contents, and is not so likely to be over-loaded; and the increased quantity of saliva thrown out in the protracted maceration of the food, softens it, and makes it more fit for digestion.

Cut feed may be composed of equal quantities of clover or of meadow hay; and wheaten, oaten, or barley straw, cut into pieces of a half or an inch in length, and mingled well together; the allowance of oats or beans is afterwards added, and mixed with the whole. Many farmers very properly bruise the oats or beans. The whole oat is apt to slip out of the feed and be lost; but when it is bruised, and especially if the feed is wet a little, it will not readily separate, or, should a portion of it escape the grinders, it will be partly prepared for digestion by the act of bruising. The prejudice against bruising the oats is utterly unfounded, so far as the farmer's horse, and the wagon horse, and every horse of slow draught, are concerned. The quantity of straw in the feed will always counteract, any supposed purgative quality in bruised oats. Horses of quicker draught, unless they are actually inclined to scour, will thrive better on bruised than on whole oats; for a greater quantity of nutriment will be extracted from the food, and it will always be easy to apportion the quantity of straw or beans to the effect of the mixture on the bowels of the horse. The principal alteration that should be made for the horse of harder and more rapid work, such as the post-horse and the stage-coach horse, is to increase the quantity of hay, and diminish that of straw. Two trusses of hay may be cut with one of straw.

Some gentlemen, in defiance of the prejudice and opposition of the coachman or groom, have introduced this mode of feeding into the stables of their horses, and with manifest advantage. There has been no loss of condition or power, and considerable saving of provender. This system is not however, calculated for the hunter, or the race-horse. Their food must lie in smaller bulk, in order that the action of the lungs may not be impeded by the distention of the stomach; yet many hunters have gone well over the field who have been manger-fed, the proportion of grain, however, being materially increased.

For the agricultural and cart-horse, eight pounds of oats, and two of beans should be added to every twenty pounds of cut feed. Thirty-four or thirty-six pounds of the mixture will be sufficient for any horse of moderate size, with fair, or even hard, work. The dray and wagon horse may require forty pounds. Hay in the rack at night is, in this case, supposed to be altogether omitted. The rack, however, may remain, as occasionally useful for the sick horse, or to contain green feed.

Horses are very fond of this provender. The great majority of them, after having become accustomed to it, will leave the best oats given to them alone, for the sake of the mingled cut feed and grain. The farmer should be cautioned, however, not to set apart damaged hay for the manufacture of the cut feed. The horse may thus be induced to eat that which he would otherwise refuse, and if the nourishing property of the hay has been impaired, or it has acquired an injurious principle, the animal will either lose condition, or become diseased. Much more injury is done by eating damaged hay, or musty oats, than is generally imagined. There will be sufficient saving in the diminished cost of the provender by the introduction of the straw and the improved condition of the horse, without poisoning him with the refuse of the farm. For old horses, and for those with defective teeth, cut feed is peculiarly useful, and for them the grain should be broken down as well as the fodder.

While the mixture of the cut feed with the grain prevents it from being too rapidly devoured and a portion of it swallowed whole, and therefore the stomach is not too loaded with that on which, as containing the most nutriment, its chief digestive power should be exerted; yet, on the whole, a great deal of time is gained by this mode of feeding, and more is left for rest. When a horse comes in wearied at the close of the day, it occupies, after he has eaten his grain, two or three hours to clear his rack. On the system of manger-feeding, the chaff being already cut into small pieces, and the beans and oats bruised, he is able fully to satisfy his appetite in an hour and a half. Two additional hours are therefore devoted to rest. This is a circumstance deserving of much consideration, even in the farmer's stable; and of immense consequence to the stage-coach proprietor, the livery-stable keeper, and the owner of every hard-worked horse.

Manger food will be the usual support of the farmer's horse during the winter, and while at constant or occasional hard work; but from the middle of April to the end of July, he may be fed with this mixture in the day, and turned out at night, or he may remain out during every rest-day. A team in constant employ should not, however, be suffered to be out at night after the end of July.

The farmer should take care that the pasture is thick and good; and that the distance from the yard is not too great, or the fields too large, otherwise a very considerable portion of time will be occupied in catching the horse in the morning. He will likewise have to take into consideration the sale he would have for his hay, and the necessity of sweet and untrodden pasture for his cattle. On the whole, however, turning out in this way, when circumstances will admit of it, will be found to be more beneficial for the horse, and cheaper than soiling in the yard.

OUT TO GRASS.

The horse of the inferior farmer is sometimes fed on hay or grass alone, and the animal, although he rarely gets a feed of grain, maintains himself in tolerable condition, and performs the work required of him; but hay and grass alone however good in quality, or in whatever quantity allowed, will not support a horse under hard work. Other substances, containing a large proportion of nutriment in a smaller compass, have been added; a brief enumeration of which follows, and an estimate is formed of their comparative value.

In almost every part of Great Britain and this country, OATS have been selected as that portion of the food which is to afford the principal nourishment. They contain from seven hundred and forty-three to seven hundred and fifty parts of the nutritive matter. They should be about, or somewhat less than, a year old, heavy, dry, and sweet. New oats will weigh ten or fifteen per cent. more than old ones; but the difference consists principally in watery matter, which is gradually evaporated. New oats are not so readily ground down by the teeth as old ones. They form a more glutinous mass, difficult to digest, and, when eaten in considerable quantities, are apt to occasion colic, or even staggers. If they are to be used before they are from three to five months old, they would be materially improved by a little kiln-drying. There is no fear for the horses from simple drying, if the grain is good when put into the kiln. The old oat forms, when chewed, a smooth and uniform mass, which readily dissolves in the stomach, and yields the nourishment which it contains. Perhaps some chemical change may have been slowly effected in the old oat, disposing it to be more readily assimilated. Oats should be plump, bright in color, and free from unpleasant smell or taste. The musty smell of wet or damaged grain is produced by a fungus growing upon the seed, which has an injurious effect upon the urinary organs, and often on the intestines, producing profuse staling, inflammation of the kidneys, colic, and inflammation of the bowels.

This musty smell is removed by kiln-drying the oat; but care is here requisite that too great a degree of heat is not employed. It should be sufficient to destroy the fungus without injuring the life of the seed. A considerable improvement would be effected by cutting the unthrashed oat-straw into chaff, and the expense of thrashing would be saved. Oat-straw is better than that of barley, but does not contain so much nutriment as that of wheat.

When the horse is fed on hay and oats, the quantity of the oats must vary with his size and the work to be performed. In winter, four feeds, or from ten to fourteen pounds of oats in the day, will be a fair allowance for a horse of fifteen hands and one or two inches in height, and that has moderate work.

In summer, half the quantity with green feed will be sufficient. Those which work on the farm have from ten to fourteen pounds, and the hunter from twelve to sixteen. There are no efficient and safe substitutes for good oats; but, on the contrary, it may be safely asserted, that they possess an invigorating property which is found in no other kind of food.

Oatmeal forms a poultice more stimulating than one composed of linseed-meal alone—or they may be mingled in different proportions, as circumstances require. In the form of gruel, it constitutes one of the most important articles of diet for the sick horse; not, indeed, to be forced upon him, but a pail containing it being slung in his box, of which he will soon begin to drink when water is denied. Gruel is generally either not boiled long enough, or a sufficient quantity of oatmeal is not used for it. The proportions should be, a pound of meal thrown into a gallon of water, and kept constantly stirred until it boils, and five minutes afterwards.

White-water, made by stirring a pint of oatmeal in a pail of water, the chill being taken from it, is an excellent beverage for the thirsty and tired horse.

BARLEY is a common food of the horse in various parts of the continent, and, until the introduction of oats, seems to have constituted almost his only food. It is more nutritious than oats, containing nine hundred and twenty parts of nutritive matter in every thousand. There seems, however to be something necessary besides a great proportion of nutritive matter, in order to render any substance wholesome, strengthening, or fattening; therefore it is, that with many horses that are hardly worked, and, indeed, with horses generally, barley does not agree so well as oats. They are occasionally subject to inflammatory complaints, and particularly to surfeit and mange.

When barley is given, the quantity should not exceed a peck daily. It should always be bruised, and the chaff should consist of equal quantities of hay and barley-straw; and not cut too short. If the farmer has a quantity of spotted or unsalable barley that he wishes thus to get rid of, he must accustom his horses to it very gradually, or he will probably produce serious illness among them. For horses that are recovering from illness, barley, in the form of malt, is often serviceable, as tempting the appetite and recruiting the strength. It is best given in mashes —water, considerably below the boiling heat, being poured upon it, and the vessel or pail kept covered for half an hour.

Grain, fresh from the mash-tub, either alone or mixed with oats or chaff, or both, may be given occasionally to horses of slow draught; they would, however, afford very insufficient nourishment for horses of quicker or harder work.

WHEAT is more rarely given than barley. It contains nine hundred and fifty-five parts of nutritive matter. When farmers have a damaged or unmarketable sample of wheat, they sometimes give it to their horses, and, it being at first used in small quantities, they become accustomed to it, and thrive and work well; it should, however, always be bruised, and given in chaff. Wheat contains a greater portion of *gluten*, or sticky, adhesive matter, than any other kind of grain. It is difficult of digestion, and apt to cake and form obstructions in the bowels. This will more often be the case, if the horse is suffered to drink much water soon after feeding upon it.

Fermentation, colic, and death, are occasionally the consequence of eating any great quantity of wheat. A horse that is fed on it, should have very little hay. The proportion should not be more than one truss of hay to two of straw. Wheat or flour, boiled in water, to the thickness of starch, is given with good effect in over-purging, especially if combined with chalk and opium.

BRAN, or the ground husk of the wheat, used to be frequently given to sick horses, on account of the supposed advantage derived from its relaxing the bowels. There is no doubt that it does operate gently on the intestinal canal, and assists in quickening the passage of its contents, when occasionally given; but it must not be a constant, or even frequent food. Bran or pollard often accumulates in the intestines, when given injudiciously, seriously impairing the digestive powers. Bran may, however, be useful as an occasional aperient in the form of a mash, but never should become a regular article of food.

BEANS afford a striking illustration of the principle, that the nourishing or strengthening effects of the different articles of food depend more upon some peculiar property which they possess, or upon some combination which they form, than upon the actual quantity of nutritive matter. Beans contain but from five hundred and twenty to six hundred parts of nutritive matter; yet they add materially to the vigor of the horse. There are many horses that will not stand hard work without beans being mingled with their food; and there are horses, whose tendency to purge it may be necessary to restrain by the astrin-

THE AMERICAN RACER, BLACK MARIA.

gency of the bean. There are few travelers who are not aware of the difference in the spirit and continuance of the horse, whether he is allowed or denied beans during the continuance of the journey. They afford not merely a temporary stimulus, but they may be daily used without losing their power, or producing exhaustion. They are indispensable to the hard-worked coach-horse. Weakly horses could never get through their work without them; and old horses would otherwise often sink under the task imposed upon them. They should not be given whole, or split, but crushed. This will make a material difference in the quantity of nutriment which will be extracted. They are sometimes given to turf-horses, but only as an occasional stimulant. Two pounds of beans may, with advantage, be mixed with the chaff of the agricultural horse, during the winter. In summer, the quantity of beans should be lessened, or they should be altogether discontinued. Beans are generally given whole. This is very absurd; for the young horse, whose teeth are strong, seldom requires them; while the old horse, to whom they are in a measure necessary, is scarcely able to masticate them, swallows many of them which he is unable to break, and drops much grain from his mouth in the ineffectual attempt to crush them. Beans should not be merely split, but crushed; as they will even then furnish sufficient employment for the grinders of the animal. Some persons use chaff with beans, instead of oats. This may possibly be allowed with hardly-worked horses; but, in general cases, beans without oats would be too binding and stimulating, and would produce costiveness, and probably megrims or staggers.

Beans should be at least a twelvemonth old before they are given to the horse, and they should be carefully preserved from damp and mouldiness, which at least disgust the animal, if they do no other harm, and harbor an insect which destroys the inner part of the bean.

The straw of the bean is nutritive and wholesome, and is usually given to the horses. Its nutritive properties are supposed to be little inferior to those of oats. The small and plump bean is generally the best.

PEAS are occasionally given. They appear to be in a slight degree more nourishing than beans, and not so heating. They contain five hundred and seventy-four parts of nutritive matter. For horses of slow work they may be used; but the quantity of chaff should be increased, and a few oats added. They have not been found to answer with horses of quick draught. It is essential that they should be crushed; otherwise, on account of their globular form, they are apt to escape from the teeth, and many are swallowed whole. Exposed to warmth and moisture in the stomach, they swell considerably, and may painfully and injudiciously distend it. The peas that are given to horses should be sound, and at least a year old. In some sections, pea-meal is frequently used, not only as an excellent food for the horse, but as a remedy for diabetes.

LINSEED is sometimes given to sick horses—raw, ground, and boiled. It is supposed to be useful in cases of catarrh.

INDIAN CORN in combination with roots, forms a valuable article of diet. Horses will eat the mess with an avidity of appetite calculated to excite surprise at first. The mess, to which a little salt should invariably be added, will keep them in fair average condition; and those which it is desirable to fatten may have a small quantity of oats, pea or bran meal added.

Hay is most in perfection when it is about a year old. The horse, perhaps, would prefer it earlier, but it is then neither so wholesome nor so nutritive, and often has a purgative quality. When it is about a year old, it retains, or should retain, somewhat of its green color, its agreeable smell, and its pleasant taste. It has undergone the slow process of fermentation, by which the sugar which it contains is developed, and its nutritive quality is fully exercised. Old hay becomes dry and tasteless, and innutritive and unwholesome. After the grass is cut, and the hay stacked, a slight degree of fermentation takes place in it. This is necessary for the development of the saccharine principle; but it occasionally proceeds too far, and the hay becomes mow-burnt, in which state it is injurious, or even poisonous. The horse soon shows the effect which it has upon him. He has diabetes to a considerable degree; he becomes, hidebound; his strength is wasted; his thirst is excessive; and he is almost worthless.

Where the system of manger-feeding is not adopted, or where hay is still allowed at night, and chaff and grain in the day, there is no error into which the farmer is so apt to fall as to give an undue quantity, and that generally of the worst kind. The pernicious results of this practice have been already mentioned in the commencement of this head, and the practice cannot be too strongly reprobated.

It is a good practice to sprinkle the hay with water in which salt has been dissolved. It is evidently more palatable to the animal who will leave the best unsalted hay for that of an inferior quality which has been moistened with brine; and there can be no doubt that the salting materially assists the process of digestion. The preferable way of salting hay is to sprinkle it over the different layers as it is put away, or as the stack is formed. From its attraction to water, it would combine with that excess of moisture which in wet seasons, is the cause of too rapid and violent fermentation, and of the hay becoming moistened, or of the stack catching fire, and it would become more incorporated with the hay. The only objection to its being thus used is, that the color of the hay is not so bright; but this will be of little consequence for home consumption.

CLOVER is useful for soiling the horse; and clover hay is preferable to meadow hay for chaff. It will sometimes tempt the sick horse, and may be given with advantage to those of slow and heavy work; but custom seems properly to have forbidden it to the roadster or those used for quick work.

THE SWEDISH TURNIP is an article of food, the value of which, particularly for agricultural horses, has not been sufficiently appreciated. Although it is far from containing the amount of nutritive matter which many have supposed, that which it has seems to be capable of complete and easy digestion. It should be sliced with chopped straw, and without hay. It quickly fattens the horse, and produces a smooth glossy coat and a loose skin. It is a good plan to give it once a day, and that at night when the work is done.

The virtues of CARROTS are not sufficiently known, both as contributing to the strength and endurance of the sound horse, and to the rapid recovery of the sick one. To the healthy horse they should be given sliced in his chaff. Half a bushel will be a fair daily allowance. There is little provender, of which the horse is more fond. There is none better, nor, perhaps, so good. When first given, it is slightly diuretic and laxative, but as the horse becomes accustomed to it, these effects cease to be produced. They also improve the state of the skin. They form a good substitute for grass, and an excellent alterative for horses out of condition. For sick and idle horses they render grain unnecessary. They are beneficial in all chronic diseases connected with breathing, and have a marked influence upon chronic cough and broken wind. They are serviceable in diseases of the skin, and in combination with oats they restore a worn horse much sooner than oats alone.

POTATOES have been given and with advantage in their raw state, sliced with chaff; but, where it has been convenient to boil or steam them, the benefit has been far more evident. Purging then has rarely ensued. Some have given boiled potatoes alone, and horses, instead of rejecting them, have soon preferred them even to oats; but it is better to mix them with the usual manger feed, in the proportion of one pound of potatoes to two and a half pounds of the other ingredients. The use of the potato must depend upon its cheapness, and the facility for boiling it. Those who have tried potatoes extensively in the feeding of horses, assert that an acre of potatoes goes as far as four acres of hay. A horse fed upon them should have his quantity of water materially curtailed. Half a dozen horses would soon repay the expense of a steaming boiler for potatoes in the saving of provender alone, without taking into account their improved condition and capability for work.

The times of feeding should be as equally divided as convenience will permit; and when it is likely that the horse will be kept longer than usual from home, the nose-bag should invariably be taken. The small stomach of the horse is emptied in a few hours; and if he is allowed to remain hungry much beyond his accustomed time, he will afterwards devour his food so voraciously as to distend the stomach and endanger an attack of the staggers.

When extra work is required from the animal, the system of management is often injudicious; for a double feed is put upon him, and as soon as he has swallowed it, he is started. It would be far better to give him a double feed on the previous evening, which would be digested before he is wanted, and then he might set out in the morning, after a very small portion of grain had been given to him, or, perhaps, only a little hay. One of the most successful methods of enabling a horse to get well through a long journey, is to give him only a little at a time while on the road, and at night to indulge him with a double feed of grain and a full allowance of beans.

WATER.

The watering of the horse is a very important but disregarded portion of his general management, especially by the farmer. He lets his horses loose morning and night, and they go to the nearest pond or brook and drink their fill, and no harm results; for they obtain that kind of water which nature designed them to have, in a manner prepared for them by some unknown influence of the atmosphere, as well as by the deposition of many saline admixtures.

The kind of water fitted for the horse has not been, as a general thing, sufficiently considered. The difference between what is termed *hard* and *soft* water, is a circumstance of general observation. The former contains certain saline principles, which decompose some bodies, as appears in the curdling of soap, and prevent the decomposition of others, as in the making of tea, the boiling of vegetables, and the process of brewing.

It is natural to suppose that these different kinds of water would produce somewhat differing effects upon the animal frame: and such is the case. Hard water, freshly drawn from the well, will frequently roughen the coat of the horse unaccustomed to it, or cause griping pains, or materially lessen the animal's power of exertion. The racing and the hunting-groom are perfectly aware of this; and instinct or experience has made even the horse conscious of it, for he will never drink hard water if he has access to soft, and he will leave the most transparent and the purest water of the well for a river, although the stream may be turbid, and even for the muddiest pool. Some trainers, indeed, have so much fear of hard or strange water, that they carry with them to the different courses the water which the animal has been accustomed to drink, and that which they know agrees with it.

The temperature of the water is of far more consequence than its hardness. It will rarely harm if taken from the pond or the running stream; but its coldness, when recently drawn from the well, has often proved injurious; it has produced colic, spasms, and even death.

There is often considerable prejudice against the horse being fairly supplied with water. It is supposed to chill him, to injure his wind, or to incapacitate him for hard work. It certainly would do so, if, immediately after drinking his fill, he were galloped hard; but not if he were suffered to quench his thirst more frequently when at rest in the stable. The horse that has free access to water, will not drink so much in the course of the day as another, who, in order to cool his parched mouth, swallows as fast as he can, and knows not when to stop.

A horse may, with perfect safety, be far more liberally supplied with water than he generally is. An hour before his work commences, he should be permitted to drink a couple of quarts. A greater quantity might probably be objectionable. He will perform his task far more pleasantly and effectually than with a parched mouth and tormenting thirst. The prejudice both of the hunting and the training groom on this point is cruel, as well as injurious. The task or the journey being accomplished, and the horse having had his head and neck dressed, his legs and feet washed, should have his water before his body is cleaned. When dressed, his grain may be offered to him, which he will readily take; but water should never be given immediately before or after the grain.

If the horse were watered three times a day, especially in summer, he would often be saved from the sad torture of thirst and from many a disease. Whoever has observed the eagerness with which the overworked horse, hot and tired, plunges his muzzle into the pail, and the difficulty of stopping him before he has drained the last drop, may form some idea of his previous suffering, and will not wonder at the violent spasms, inflammation, and sudden death, that often follow.

It is a judicious rule with travelers, that when a horse begins to refuse his food, he should be pushed no further that day. It may, however, be worth while to ascertain whether this does not proceed from thirst as much as from exhaustion; for in many instances his appetite and his spirits will return soon after he has partaken of the refreshing draught.

PASTURING.

So far as mere health is concerned, grass is the most salubrious food which the horse can receive. When it is eaten where it grows, the horse is said to be turned out, to be getting a run at grass, or to be at grass. When it is cut, and consumed in the stable, the horse is said to be soiled.

It is probable that grass eaten in the field produces quite the same effects as that eaten in the stable. But at pasture, there are several agents in operation to which the stabled horse is not necessarily exposed. The exercise which he must take; the position which his head must assume, in order that he may obtain food; the annoyance he suffers from flies; his exposure to the weather; the influence of the soil upon the feet and legs; and the quantity of food placed at his disposal; are the principal points wherein pasturing differs from soiling.

The EXERCISE which he must take as he gathers his food, varies according to the herbage. When the ground is bare, the exercise may amount even to work, but to a sound horse it is never injurious; in cold weather it keeps him warm, or, at least, prevents him from becoming very cold. With a lame horse, the case is different. In some species of lameness, as in chronic diseases of the joints, the slow but constant exercise thus rendered necessary is highly beneficial; but the exertion demanded by a bare pasture is unfavorable to any sprain or lameness arising from disease in the ligaments and tendons. Lameness, when very great, no matter where seated, forbids pasturing, even though the grass be knee-high. The pain of standing, and moving on two or three legs, may be so great that the horse will be obliged to lie down before he has obtained half a meal. It is for slight lameness only that horses should be turned out; and the pasture should be such as to afford sufficient nutriment, without giving the horse more exercise than is good for the disease.

The legs of fast-working horses often become turned, shapeless, tottering, bent at the knee, and straight at the pasterns. These always improve at pasture, as, indeed, they do in the stable, or loose-box, when the horse is thrown out of work. Grazing exercise does not appear to be unfavorable to their restoration; but when the knees are very much bent, the horse is unfit for turning out; he cannot graze; when his head is down, he is ready to fall upon his nose, and it costs him much effort to maintain his balance.

PASTURING.

THE POSITION OF THE HEAD in the act of grazing is unfavorable to the return of blood from the brain, from the eyes, from all parts of the head. Horses that have had staggers, or bad eyes, those that have recently lost a jugular vein, and those that have any disease about the head—strangles, for instance— should not be sent to pasture. The disease becomes worse, or, if gone, is apt to return. Even healthy horses are liable to attacks on the brain, when turned to grass, particularly when the weather is hot, and the herbage abundant.

Horses that have been for more than a year in the stable,

and especially those that have been reined up in harness, often experience considerable difficulty in grazing. The neck is rigid, and the muscles which support the head are short. It is often several weeks before an old coach-horse can graze with ease. Very old coach-horses that have short, stiff necks, should not be turned out when they can be kept in; if they must go, they should be watched, lest they die of want.

EXPOSURE TO THE WEATHER. Wet, cold weather always produces emaciation and a long coat. If the horse is put out without preparation, he is apt to have an attack of inflamed lungs, or sore throat, or a common cold, with discharge from the nose, and may sicken and die. Many persons seem to think that no usage is too bad for the horse, if it do not immediately produce some fatal disease. Early in spring, or late in autumn, the animal is turned out of a warm, comfortable stable, and left to battle with the weather as he best can. He crouches to the side of a wall, shivering and neglected, as if he had no friend in the world. In time, the horse becomes inured to the weather, if he does not sink under it, but sometimes he comes home with diseased lungs, and very often with a cough which never leaves him, and which produces broken wind.

SHELTER, so easily provided—at the cost of a few rude boards even—is too much neglected in the pasture. A hovel, covered on three sides, the fourth open to the south, and just high enough to admit the horse, will answer the purpose. The bottom should be sloping, elevated, and quite dry. When litter can be afforded, it will tempt the horse out of the blast. There may be hay-racks and mangers, strong, though of rude construction. In summer, the horse can retire here during the heat of the day, and in the more inclement season he may thus avoid the wind and the storm.

Exposure to hot weather is not so pernicious, although it always produces pain, if the horse be turned out in the middle of summer. For a while he is fevered all day and loses flesh; but he soon recovers. The parts that are most apt to suffer are the brain and the eyes. Staggers, that is, an affection of the brain, is not common, and the eyes never suffer permanent mischief. They are inflamed by the flies, but the brain is injured, partly by the heat, and partly by the pendent position of the head.

FLIES. The horse is persecuted by at least three kinds of flies. One, the common horse-fly, settles on his ears and different parts of his body, tickling and teazing him. Another is a large fly, termed the gad-fly; it is a blood-sucker, bites pretty smartly, and irritates some tender-skinned horses* almost to madness, forcing them sometimes to rush into the water to escape their attacks. Another fly is a small insect, whose name is unknown, which lives in the blood, attacking those parts where the skin is thinnest, as the eyelids, inside and outside, the sheath, and the vagina. The eyelids especially always swell where this fly abounds, and the swelling is sometimes so great as to make the horse nearly blind, while the eye is red and weeping. The injury however, is not permanent.

The principal defense which the horse has against these puny, but tormenting enemies, is his tail. On some parts of his body he can remove them with his teeth and his feet; and that which cannot be done by these, is done by the tail. With us, however, in far too many instances the effective instrument which nature has furnished is removed, or materially impaired, before he has attained maturity; and, as if the pains of servitude were not sufficiently great and numerous, domestication is rendered still more intolerable by whim and caprice.

THE SOIL. Much has been said about the influence of the soil upon the horse's feet and legs, and much exaggeration of assertion has been set afloat. Horses reared in soft, marshy pastures have large flat feet, low at the heels, and weak everywhere. On dry ground the hoof is hard, strong, and small, the sole concave, and the heels high. But to impart any peculiar character to the hoof, or to produce any change upon it, a long and continuous residence upon the same kind of soil is necessary. A period of six months may produce some change; but it is so insignificant in general that it is not apparent.

The low temperature at which the feet and legs are kept in a moist pasture has probably some influence, though not very great, in abating inflammation in those parts. The legs become finer and free from tumors and gourdiness; but they would improve nearly or quite as soon, and as much, in a loose box.

When the pastures are hard and baked by the sun, unshod horses are apt to break away the crust, and they often come home with hardly horn enough to hold a nail. Feet that have never been shod suffer less; others should, as a general thing, be preserved by light shoes, especially on the fore feet; kicking horses, when shod behind, are rather dangerous among others.

It has been supposed that the act of grazing throws considerable stress upon the tendons of the fore legs, and ultimately impairs them. This has been urged against grazing hunters; but so far as sound legs are concerned, there seems to be no foundation for the supposition, and it certainly has never been proved.

QUANTITY OF FOOD. In the stable, a horse's food can be apportioned to him as his wants may require; but at pasture, he may get too much or too little. It is difficult to put the horse where he will obtain all the nourishment he needs, and no more. In a rich pasture, he may acquire an inconvenient load of fat; in a poor one, he may be half starved. If he must go out, he may be taken in before he becomes too fat; or he may be placed in a bad pasture, and fed up to the point required by a daily allowance of grain.

TIME OF TURNING OUT. Horses are pastured at all times of the year. Some are out for lameness, some for bad health, and some, that they may be kept for less than the stable cost. The usual time of turning out is about the end of April, or the beginning of May. Then the grass is young, juicy, tender, and more laxative than at a later period. The spring grass is best for a horse in bad health, worn out by sickness, hard work, or bad food. The weather is mild, neither too hot nor too cold; when it is unsettled and backward, the delicate horse, and sometimes every one, should come in at night and on bleak days. Toward the end of summer, the grass is hard, dry, coarse, fit enough to afford nutriment, but not to renovate a shattered constitution. The days are hot, the nights cold and damp, and the flies strong and numerous. This is not the time for turning out a delicate or thin-skinned horse.

Many persons are accustomed to give the horse a dose or two of physic before sending him to grass. Unless the animal has tumid legs, or is afflicted with some ailment, this is entirely unnecessary, though it may do no harm. To prepare the horse for exposure to the weather, the clothing to which he has been accustomed is lightened, and then entirely removed, a week or two before turning him out. The temperature of the stable is gradually reduced, until it becomes as cool as the external air. These precautions are most necessary for horses that have been much in the stable, and particularly a warm stable. For eight or ten days previous to going out, the animal should not be groomed. The dust and perspiration which accumulate upon the hair, seem in some measure to protect the skin from rain and from flies. The feet should be dressed, and the grass shoes, or plates, applied a week before turning out. If they are injured by the nails, the injury will become apparent before much mischief is done; at grass it might not be noticed so soon. On the day of going out, the horse should be fed as usual. If he goes to grass when very hungry, he may eat too much. Indigestion will be the result, which may prove fatal. Weather permitting, night is usually chosen for the time of turning out, as the horse is not so apt to gallop about. Let loose in the day time, many are disposed to gallop till they lame themselves, and to try the fences.

In autumn, or early in spring, the stable preparation for grass is often insufficient. If the horse be tender, or the weather unsettled, he should be taken home every night, for perhaps the first week. For eight or ten days longer, it may be proper to house him on very wet or stormy nights. The stable given to him should always be cool, not so cold as the external air, but never so warm as if he were accustomed to it.

CONFINEMENT. Some horses are not easily confined at pasture. They break or leap the fences, and wander over the country, or proceed to the stable. The fore feet are sometimes shackled in order to confine them; but these fetters, if worn for a long time, are apt to alter the horse's action, rendering it short, confined, irregular, at least for a time, till he regains the use of his shoulders. Sometimes the horse is tied by a rope to a stake driven in the ground. He then requires almost constant watching, for he must be often shifted as he eats down the grass, and he may get his legs entangled in the rope, thereby casting himself, and receiving serious injury, unless relief be immediate. Sometimes he is tied to a stake, which he can drag about the field. He soon finds that he can walk where he pleases, but he cannot run, and seldom attempts to leap. This, however, is also liable to throw the horse down, or to injure his legs by getting them entangled in the rope. To prevent the horse from leaping, a board is sometimes suspended round his neck, reaching to his knees, which it as apt to bruise. None of these clumsy and unsafe restraints should ever be employed, when it is possible to dispense with them. Few horses, mares in spring and stallions excepted, require them after the first two days. For horses that are turned out only an hour or two during the day, they are as much used to enable him to be easily caught when wanted, as to prevent him from wandering.

ATTENDANCE WHILE OUT. Horses at grass should be visited at least once every day. If neglected for weeks, as often happens, one may be stolen, and conveyed out of the country before he is missed; the fences may be broken; the water may fail; the horses may be lamed, or attacked with sickness; one may roll into a ditch, and die there for want of assistance to extricate him; the shoes may be cast; the heels may crack;

thrushes may form; sores may run into sinuses, or become full of maggots; the feet and legs may be injured by stubs, thorns, broken glass, or kicks; or the horses may quarrel, fight, and wound each other. That these and similar evils and accidents may be obviated, or soon repaired, the horses should be visited every morning by a trustworthy person who knows what is required of him.

The grain, hay—either or both—if any be given, should be furnished at regular intervals; when fed with grain, the horses ought to be watched till it is eaten, lest they rob each other, or some prowling thief appropriate the whole. Horses at grass require, and should have, no dressing, as it exposes the skin too much. The shoes may be removed, however, and the feet dressed every four or five weeks.

TREATMENT AFTER GRAZING. When taken from grass to warm stables, and put upon rich, constipating food, horses frequently become diseased. Some catch cold, some suffer inflammation in the eyes, some take swelled legs, cracked heels, grease, thrushes, founders, surfeit, or a kind of mange. These are very common; and physic is often, and indeed generally, given to prevent them. They are produced by a combination of circumstances; by sudden transition from gentle exercise and indolence or exciting work; from a temperate to stimulating diet; from a pure, cool, and moving atmosphere, to an air comparatively corrupt, hot, and stagnant. These changes must be made, and are, to a certain extent, unavoidable; but it is not in all cases necessary that they should be made suddenly. It is the rapid transition from one thing to another and a different thing, that does all the mischief. If it were effected by slow degrees, the evils would be avoided, and there would be less need, or none at all, for those medicines which are given to prevent them.

During the first week, the temperature of the stable ought to be little different from that of the external air. Subsequently it may be raised, by slow degrees, till it is as warm as the work or other circumstances demand. The horse should not at first be clothed, and his first clothing should be light. Grooming may commence on the first day; but it is not good to expose the skin very quickly by a thorough dressing. The food should be laxative, consisting of bran-mashes, oats, and hay; but no beans, or very few. Walking-exercise, twice a day, is absolutely necessary for keeping the legs clean, and it assists materially in preventing plethora.

The time required for inuring a horse to stable treatment, depends upon several circumstances. If taken home in warm weather, the innovation, so far as the temperance and the purity of the air are concerned, may be completed in about two weeks. If the horse is not very lean, his skin may be well cleaned in the first week; and to clean it, he must have one or two gentle sweats, sufficient to detach and dissolve the dust, mud, and oily matter which adhere to the skin, and glue the hair together. All this, or as much of it as possible, must be scraped off while the horse is warm and perspiring. If it is allowed to get dry before scraping, he is just where he was. If the weather be cold, there need be no great hurry about cleaning him completely.

The propriety of giving physic after grazing has been often questioned. In the stable, its utility is generally acknowledged. In books it is sometimes condemned as pernicious, sometimes as useless. It may be safely said, however, that there are many cases in which physic is very useful; but that as a general thing, it is given too indiscriminately, and before it is wanted.

To a lusty horse, one or two doses may be given for the purpose of reducing him, for removing superfluous fat and flesh. The physic may be strong, sufficiently so to produce copious purgation. It empties the bowels, takes up the carcass, and gives freedom to respiration; it promotes absorption, and expels the juices which embarrass exertion. Work, sweating and a spare diet of condensed food, will produce effects without the aid of physic. But purgation shortens the time of training, and it saves the legs. If the horse must be rapidly prepared for work, with as little hazard as possible to his legs, he must have physic. The first dose may be given on the day when he comes from grass; the others, if more than one be necessary, at intervals of eight or ten clear days.

A lean horse, fresh from grass, needs no physic till he has been stabled for several days, and perhaps not then. By the time he has acquired strength sufficient to stand training, his bowels are void of grass, and his belly small enough to allow freedom of respiration. At the end of a fortnight or three weeks, the lean horse ought to be decidedly lustier. If too much so, and acquiring flesh too rapidly, one dose of physic may be given, active enough to produce smart purgation, and prevent the evils which arise from plethora. If he is not taking on flesh so rapidly as he should, he may have two, perhaps three, mild doses of physic, just active enough to produce one or two watery or semi-fluid evacuations. If he eat a great deal without improving in condition, he is probably **troubled with worms,** and half a drachm of calomel may be added to each dose of physic. If he does not feed well, there is probably a torpid state of the digestive apparatus, produced by a bad or deficient diet. In such a case, mild physic is still proper, and, in addition, the horse may have a few tonic balls between the setting of one dose and the administration of another. Four drachms of gentian, two of ginger, and one of tartar emetic, made into a ball with honey, forms a very useful tonic. One of these may be given every day, or every second day, for a fortnight. If the horse does not improve under these, he requires the aid of a veterinary surgeon.

The mode of grazing farm-horses requires some notice. Other horses are sent to pasture, and with few exceptions, remain at it for days and weeks without interruption. Those employed in agriculture are pastured in three different ways. By one, the horse is constantly at grass, except during his hours of work; he is put out at night, is brought in the next morning, goes to work for two or three hours, and is then returned to pasture for about two hours; in the afternoon he again goes to work, which may be concluded at five or six o'clock, and from that time till he is wanted on the next morning he is kept at grass. By another mode, the horse is turned out only at night. During the day he is soiled in his stable at his resting intervals. When work is over for the day, he is sent out till the next morning. By the third mode, which is generally allowed to be the best, the horse is turned to grass only once a week. He is pastured from the time his work is finished on Saturday night till it commences again on Monday morning.

If the horses have any thing like work, the first two modes are decidedly objectionable. There is much expenditure of labor in procuring the food, and there is great loss of time. It may cost the horse four or five hours good work to cut down the grass which he eats. A man supplied with a scythe will do the same work with far less labor in a few minutes. If there be nothing else for the horse to do, it is quite right to make him gather his own food. But, otherwise, it is absurd to make him exhaust his strength and time in doing that which a man can do so much more easily and quickly. Besides this expenditure of the horse's time and strength, the loss of manure, and the damage done to pasture by the feet, ought to be taken into consideration.

The third mode of grazing appears to be the least objectionable. The horses have no field labor on Sunday; if the pasture be good, the weather favorable, and the horses not fatigued, they are better at grass than in the house.

In some places the road-horses are sometimes put to grass on Sunday. This practice has nothing apparently to recommend it. The weekly work of these horses in general demands the rest which Sunday brings; and if they travel at a fast pace, as all coach-horses do now, they are apt to eat so much grass, and carry such a load in their bellies, that on Monday they are easily over-worked. The breathing is impeded, unless the horses purge, which few do. They often come from grass as haggard and dejected as if they had done twice their ordinary work the day before.

SERVICE.

A change of lodging, or of diet, is often a cause of disease. When a fresh horse is procured, it is well to know how he has been treated during the previous month; if he is a valuable animal, he will certainly be worth this inquiry. Horses that come from a dealer have probably been standing in a warm stable, well-clothed, well-groomed, highly fed, and seldom exercised. They have fine glossy coats, are lusty, and in high spirits; but their flesh is soft and flabby. They are unfit for fast work; they are easily heated by exertion, and when the least warm,

SERVICE.

are very apt to take cold. But, wherever the horse comes from, or whatever his condition may be, changes in reference to food, temperature, and work, must be effected by slow degrees. It is absurd and always pernicious to take a horse from the field, and put him in a warm stable, and on rich food all at once; it is no less erroneous to take him from a warm to a cold stable, or to demand exertion to which he has not been trained.

When the horse's history cannot be traced, both his work and his diet should at first be moderate. More of either than he has been accustomed to, will do more harm than less of either. It may, however, soon be ascertained by trying him whether he has been doing much work; if fit for work, he may be fed in proportion. The temperature of the stable had better be warmer than colder. If too warm, the horse will perspire; his **coat will be wet in different places, especially in the morning** when the stables are first opened. If it be too cold, his coat will be roughened, and become dim, and the horse will catch cold, evidence of which will be given by a cough.

The work of some horses exposes them much to the weather. Those employed in street-coaches, in the carriages of medical men, all those that have to stand in the weather, can never do so with safety until they have been seasoned. In the cold rainy season, many are destroyed, and many more endangered by injudicious exposure. Wet weather is the most pernicious; yet it is not the rain alone that does the mischief. If the horse is kept in motion, and afterwards perfectly and quickly dried, or is kept in motion till he is dry, he suffers no injury. His coat may be bleached till it is like a dead fur; but the horse does not catch cold. If he is allowed to stand at rest with his coat drenched in the rain, the surface of the body rapidly loses its heat, there being no stimulus to the formation of it; the blood circulates slowly, accumulates internally, and oppresses vital organs, especially the lungs; the legs become excessively cold and benumbed; the horse can hardly use them, and, when put in motion, he strikes one against the other. Exposure, when it deprives the body of heat in this way, is a fruitful source of inflamed lungs, of thoracic influenza, catarrh, and founder. When the skin is wet, or the air very cold, the horse should, if possible, be kept in motion, which will preserve him, however little he may have been accustomed to exposure.

Horses that have been kept in warm stables, and never out but in fair weather, are in most danger. If they cannot be kept in constant motion, they must be prepared before they are exposed. If they commence work in summer, or early in the autumn, they will be fully inured to the weather before the worst part of winter arrives. But if they commence in winter, they should be out for only one or two hours at a time; in good days they may be out longer, no one being able to give a precise rule as to the length of time appropriate, as it varies with the condition of the animal, the weather, and the work required. It should shorten with the wetness or coldness of the weather, and the tenderness of the animal. If he must run rapidly from one place to another, and wait perhaps half an hour at each, he is in more danger than if the pace were slower, and the time of waiting shorter; and if moved about constantly, or every ten minutes, he suffers less injury than if he was standing still. After a time he becomes inured to exposure, and may be safely trusted in the severest weather.

Repeated and continued application of cold to the surface of the body stimulates the skin to produce an extra supply of heat. The exposure of two or three days is not sufficient to rouse the skin to this effort. It is always throwing off a large quantity of heat; but it is several days, and with many horses several weeks, before the skin can assume activity sufficient to meet the demands of a cold or wet atmosphere. Ultimately, it becomes so vigorous that the application of cold, whether wet or dry, is almost instantly followed by an increased production of heat. To this, however, there are limits. By exposure, gradually increasing in length and frequency, the system may be able to maintain the temperature at a comfortable warmth for three or four successive hours, even when the horse is standing at rest in wet or cold. But he cannot endure this beyond a certain point. Exhaustion and emaciation succeed, in spite of all the food the horse can eat. The formation of so much heat consumes the nutriment that ought to produce vigor for work.

Hence, working horses kept very much in very cold stables are lean and dull.

It is chiefly the horses that have to *stand* in the weather, which require preparation for exposure. Bleeding, purging, and other means, which debilitate or emaciate, are never necessary in this process. Hunting, stage-coach, and cart-horses seldom require any preparation for exposure, as they are in motion from the time of leaving the stable till their return. They only require to be well and quickly dried when wet.

New horses are very liable to have the skin injured by the harness. The friction of the saddle, collar, or traces, produces excoriation. In some horses this is altogether unavoidable, especially when they are in poor condition. Their skin is tender, and a little chafing exposes the quick. In all horses it is some time before the skin thickens, and becomes sufficiently callous to carry the harness without injury. The time required to undergo this change varies materially, and cannot be much shortened by any means. Attention to the harness, however, will frequently prevent excoriation. After every journey, the neck should be closely examined. If there be any spot, however little abraded, hot and tender when pinched, that part of the collar which produced it should be cut out before the next journey. The guard, or safe, is a useful article to prevent galls of this kind. It is merely a thin slip of soft leather, covering the seat of the collar. It obviates friction, and prevents injurious pressure from any little protuberance or hardness in the stuffing of the collar. On the first or second journey a new horse often comes in with his neck somewhat inflamed; it is hot, tender, and covered with pimples. In the stables it is said to be *fired.* A solution of common salt in water is commonly applied, and it serves to allay the inflammation; it should be applied whenever the collar is removed. Tumors, containing bloody water, frequently rise on the neck. They should be opened immediately, emptied, and kept opened for a few days. The piece must be taken out of the collar, and a safe used. On a hilly road the lower part of the collar often galls the neck seriously, in spite of any alteration in the stuffing. A broad strap attached to the collar, and passing over the windpipe, is a good remedy. The strap should be two inches broad, and drawn tightly enough to keep the collar steady, and make it stand nearly upright. It should be adjusted before the head is put on the bearing reins, and should be worn till the neck is quite sound. A broad breast-band may also be substituted for the neck collar, till the neck and shoulders get well. A horse will pull nearly as well in this as in the collar and hames. When the traces, crupper, or pad, threaten or produce excoriation, they must be kept off by cushions placed behind, before, or at each side of the part injured.

The back requires nearly as much care as the neck. A new saddle is objectionable for a new horse, particularly when he has to travel far under a heavy rider. A tender back may be hardened by frequent use of the saddle and a light weight. The horse may stand saddled in the stable, and saddled when he goes to exercise. When the back is hot, and the skin disposed to rise in tumors, the saddle should remain on till the back becomes cool. Slacken the girths, raise the saddle for a moment, and then replace it. Its weight prevents tumors; excoriation and firing must be treated as on the neck. Always let the pommel of the saddle be dry before it is again used, and put it on half an hour before the horse is to be mounted.

Horses, from whom extraordinary exertions are not demanded, and those that are never expected or required to do all that a horse is capable of doing, stand in little need of inurement to work, and it is seldom that any is intentionally given. When a saddle or draught-horse is purchased, he is often put to his work at once, without any preparation. He is treated as if he were as able for the work as it is possible to make him. So long as the work is slow and not very laborious, he may perform it well enough; but this system will not do for full work, whether fast or slow. If the horse has been idle for a month or two, he is weak. It matters little that he is plump and in good spirits. He may be able to draw a load of twenty or thirty hundred-weight with ease, and perhaps to draw it a considerable distance; but on the next day he is sore all over, stiff, feeble, dull, almost unable to carry his own weight. If the same work be exacted day after day, the horse loses flesh, and at last becomes unfit for any work. But if the werk be less severe at first, and gradually increases from week to week, the horse at last acquires strength and endurance greater, perhaps, than he ever before possessed. He is then able to do with ease as much in a week as would have completely knocked him up at the beginning. For slow, moderate work, this is all the preparation which the horse needs. At first, let it be very gentle; and the weight he is to carry or draw, and the distance he is to travel, may be increased as he is found able to bear it. In preparing the horse for hunting, racing, or coaching, the treatment must be somewhat different.

SHOEING.

There is hardly any other class of mechanics who combine so much ignorance of the principles on which their art is founded, with so much conceit of their knowledge, as do ordinary horse-shoers; and it should be one of the first duties of the horse-owner to inform himself of the nature and structure of the horse's foot, the reason why shoeing is necessary at all, what parts of the foot it protects, what is the best form of shoe to effect the purpose, how it may be best fastened to the foot, and how often it should be removed.

To illustrate these important points, cuts are here introduced, showing the construction of the horse's foot.

Our first one shows the ground surface of the hoof prepared for receiving a shoe; and marks very distinctly the difference between the curvature of the outer and inner quarters.

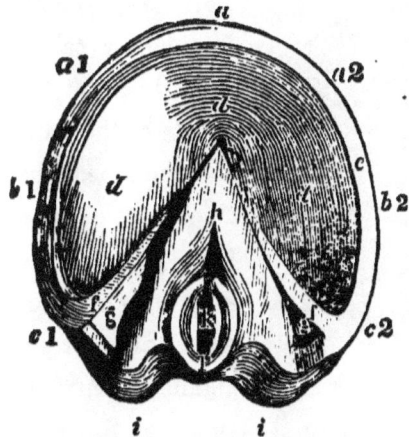

GROUND SURFACE OF THE HOOF.

The hoof is divided into horny crust or wall, sole, and frog. The horny crust is secreted by the numerous blood-vessels of that soft protruding band which encircles the upper edge of the hoof, immediately beneath the termination of the hair; and is divided into toe, quarters, heels, and bars. Its texture is insensible, but elas-

a. The toe—rasped away to receive the turned-up shoe. *a* 1. The *inner* toe. *a* 2. The *outer* toe. *b* 1. The *inner* quarter. *b* 2. The *outer* quarter. *c* 1. The *inner* heel. *c* 2. The *outer* heel. *d. d. d.* The sole. *e. e.* The crust or wall of the hoof. *f. f.* The bars. *g. g.* The commissures. *h. k. l.* The frog. *h.* The part immediately under the navicular joint. *k.* The oval cleft of the frog. *l.* The elevated boundary of the cleft. *i. i.* The bulbs of the heels.

tic throughout its whole extent; and, yielding to the weight of the horse, allows the horny sole to descend, whereby much inconvenient concussion of the internal parts of the foot is avoided. But if a large portion of the circumference of the foot is fettered by iron and nails, it is plain that that portion, at least, cannot expand as before; and the beautiful and efficient apparatus for effecting this necessary elasticity, being no longer allowed to act by reason of these restraints, becomes altered in structure; and the continued operation of the same causes, in the end, circumscribes the elasticity to those parts alone where no nails have been driven; giving rise to a train of consequences destructive to the soundness of the foot, and fatal to the usefulness of the horse.

The toe of the fore foot is the thickest and strongest portion of the hoof, and is in consequence less expansive than any other part, and therefore better calculated to resist the effects of the nails and the shoe. The thickness of the horn gradually diminishes toward the quarters and heels, particu-

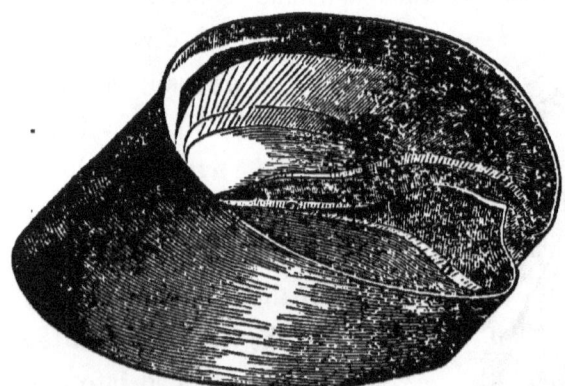

THE HOOF OF THE HORSE.

*. Is a broad flat mass of horn, projecting upward into the middle of the elastic cushion, and called "the frog stay." *b, b.* Are two horny projections rising into the cavity of the hoof formed by the commissures. *c. c.* Are portions of the same projections, and are situated just under the two ends of the navicular bone, and mark the point on either side where diminution in the natural elasticity of the fatty frog would be felt with the greatest severity by the navicular joint; for under the most favorable circumstances, the quantity of cushion between these points and the navicular joint cannot be very large; and hence the importance of our doing all we can to preserve its elasticity as long as possible.

larly on the inner side of the foot, whereby the power of yielding and expanding to the weight of the horse is proportionably increased, clearly indicating that those parts cannot be nailed to an unyielding bar of iron, without a most mischievous interference with the natural functions of the foot. In the hind foot, greater thickness of horn will be found at the quarters and heels, than in the fore foot. This difference in the thickness of horn is beautifully adapted to the inequality of the weight which each has to sustain, the force with which it is applied, and the portions of the hoof upon which it falls.

The toe of the fore foot encounters the combined force and weight of the fore hand and body, and consequently in a state of nature is exposed to considerable wear and tear, and calls for greater strength and substance of horn than is needed by any portion of the hind foot, where the duty of supporting the hinder parts alone is distributed on the quarters and heels of both sides of the foot.

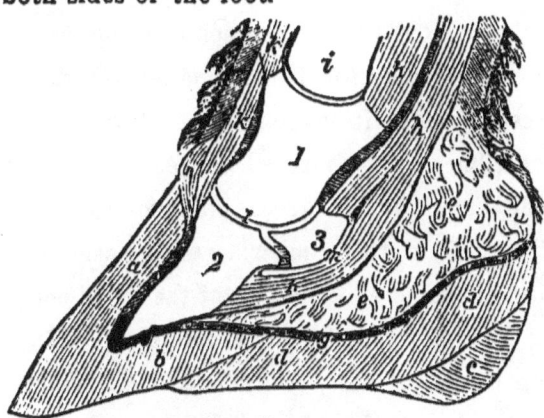

A SECTION OF THE FOOT.

The bars are continuations of the wall, reflected at the heel towards the centre of the foot, where they meet in a point, leaving a triangular space between them for the frog.

1. The coronet bone. 2. The coffin bone. 3. The navicular bone. *a.* The wall. *b.* The sole. *c.* The cleft of the frog. *d. d.* The frog. *e. e.* The fatty frog, or elastic cushion. *f.* The sensitive sole. *g.* The sensitive frog. *h. h. h.* Tendons of the muscles which bend the foot. *i.* Part of the pastern bone. *k. k.* Tendons of the muscles which extend the foot. *l.* The coffin joint. *m.* The navicular joint. *n.* The coronary substance. *o.* The sensible laminæ, or covering of the coffin bone.

The whole inner surface of the horny crust, from the centre of the toe to the point where the bars meet, is everywhere lined with innumerable narrow, thin, and projecting horny plates, which extend in a slanting direction from the upper edge of the wall to the line of junction between it and the sole, and possess great elasticity. These projecting plates are the means of greatly extending the surface of attachment of the hoof to the coffin bone, which is likewise covered by a similar arrangement of projecting plates, but of a highly vascular and sensitive character; and these, dovetailing with the horny projections above named, constitute a union combining strength and elasticity in a wonderful degree.

The horny sole covers the whole interior surface of the foot excepting the frog. In a well-formed foot it presents an arched appearance, and possesses considerable elasticity, by virtue of which it ascends and descends, as the weight above is either suddenly removed from it, or forcibly applied to it. This descending property of the sole calls for one especial consideration in directing the form of the shoe; for, if the shoe be so formed that the horny sole rests upon it, it cannot descend lower; and the sensitive sole above, becoming squeezed between the edges of the coffin bone and the horn, produces inflammation, and perhaps abscess. The effect of this squeezing of the sensitive sole is most commonly witnessed at the angle of the inner heel, where the descending heel of the coffin bone, forcibly pressing the vascular sole upon the horny sole, contuses a small blood-vessel, and produces what is called a corn, but which is, in fact, a bruise.

The horny frog occupies the greater part of the triangular space between the bars, and extends from the hindermost part of the foot to the centre of the sole, just over the point where the bars meet, but is united to them only at their upper edge; the sides remain unattached and separate, and form the channel called the commissures.

If we carefully observe the form and size in the frog in the foot of a colt of from four to five years old, at its first shoeing, and then note the changes which it undergoes as the shoeings are repeated, we shall soon be convinced that a visible departure from a state of health and nature is taking place. At first it will be found large and full, with considerable elasticity; the cleft oval in form, open, and expanding, with a continuous, well-defined, and somewhat elevated boundary; the bulbs at the heels fully developed, plump, and rounded; and the whole mass occupying about one-sixth of the circumference of the foot. By degrees the fulness and elasticity will be observed to have diminished; the bulb at the heels will shrink, and lose their plumpness; the cleft will become narrower, its oval form disappear, the back part of its boundary give way, and it will dwindle into a narrow crack, extended back between the wasted, or perhaps obliterated, bulbs, presenting only the miserable remains of a frog, such as may be seen in the feet of most horses long accustomed to be shod.

The bones proper to the foot are three in number,—viz., the coffin bone, the navicular bone, and part of the coronet bone; they are contained within the hoof, and combine to form the coffin joint; but the smallest of them, the navicular bone, is of far more importance as connected with the subject of shoeing, than either of the others; for upon the healthy condition of this bone, and the joint formed between it and the tendon, which passes under it to the coffin bone, and is called the navicular joint, **mainly depends the usefulness of the horse to man.**

This small bone, which in a horse sixteen hands high measures only two and a quarter inches in its longest diameter, three-fourths of an inch at the widest part of its shorter diameter, and half an inch in thickness in the centre, its thickest part, has the upper and under surfaces and part of one of the sides overlaid with a thin coating of gristle, and covered by a delicate secreting membrane, very liable upon the slightest injury to become inflamed; it is so placed in the foot as to be continually exposed to danger, being situated across the hoof, behind the coffin bone, and immediately under the coronet bone; whereby it is compelled to receive nearly the whole weight of the horse each time that the opposite foot is raised from the ground.

The coffin bone consists of a body and wings; and is fitted into the hoof, which it closely resembles in form. Its texture is particularly light and spongy, arising from the quantity of canals or tubes that traverse its substance in every direction, affording to numerous blood-vessels and nerves a safe passage to the sensitive and vascular parts surrounding it; while the unyielding nature of the bone effectually protects them from compression or injury, under every variety of movement of the horse.

In an unshod foot, the front and sides of the coffin bone are deeply furrowed and roughened, to secure the firmer attachment of the vascular membranous structure, by which the bone is clothed; but in the bone of a foot that has been frequently shod, the appearance is greatly changed, the furrows and roughness giving place to a comparatively smooth surface. This change is probably produced by the shoe limiting, if not destroying, the expansive power of that part of the horn to which it is nailed; whereby a change of structure in the membrane itself, as well as absorption of the attaching portions of the bone, is induced; for it is an invariable law of the animal economy not to continue to unemployed structures the same measure of efficient reparation that is extended to parts constantly engaged in performing their allotted tasks. The shoe restricts or prevents expansion; while nature, as the secret influence is called, immediately sets to work to simplify the apparatus for producing the expansion, which art has thus rendered impracticable, and substitutes for it a new structure, less finely organized, but admirably suited to the altered condition of the parts.

The wings extend from the body of the bone directly backward, and support the lateral cartilage of the foot.

The sensitive sole, or, as it is sometimes called, the fleshy sole, is about the eighth of an inch thick, and is almost entirely made up of blood-vessels and nerves; it is one of the most vascular and sensitive parts of the body, and is attached to the lower edge of the sensitive covering of the coffin bone, to the bars, and point of the frog, and also with great firmness to the whole of the arched under-surface of the coffin bone.

The sensitive frog includes not only the part corresponding to the sensitive sole, but also the peculiar spongy elastic substance which intervenes between it and the navicular joint, and fills the space between the cartilages. The proper sensitive frog is thicker, and less finely organized, than the sensitive sole, possessing fewer blood-vessels and nerves.

It is a common, but very erroneous, opinion, that the shape of the perfect foot is circular, or very nearly so. This induces most smiths to endeavor to reduce the foot to that shape as soon as possible. There are very few things in nature so little varied as the form of the ground surface of horses' feet; for whether the hoof be high-heeled and upright, or low-heeled and flat, large or small, broad or narrow, the identical form of ground-surface is maintained in each, so long as it is left entirely to nature's guidance. The outer quarter, back to the heel, is curved considerably and abruptly outward, while the inner quarter is carried back in a gradual and easy curve. The advantage of this form is so obvious, that it is strange that any interference should ever be attempted with it. The enlarged outer quarter extends the base, and increases the hold of the foot upon the ground; while the straighter inner quarter lessens the risk of striking the foot against the opposite leg.

The inclination of the front of the horny crust of the foot should be at an angle of about forty-five degrees. If the foot is much steeper than this, it is very liable to contract; while, if it is much more slanting, it constitutes what is called the "oyster shell" foot, in which there is an undue flatness of the sole, and a tendency to pumiced feet.

Before removing the old shoes, care should be taken to raise all the clinches of the nails to prevent injury to the crust, and to avoid giving pain to the horse; even after clinches are raised, if the shoes cannot be easily drawn off, those nails which seem to hold most firmly should be punched, or drawn out, that the shoe may be removed without injury to the hoof, and without weakening the nail-hold for the new shoeing.

The shoe being removed, the edge of the crust should be well rasped to remove so much of the horn as would have been worn away by the contact with the ground, had it been unshod. In no case should the rasp be used on the surface of the hoof, except to make the necessary depressions for the clinches, after the new shoe has been put on, and to shape the hoof *below* the line of the clinches of the nails. The hoof, above this line, will inevitably be injured by such treatment, which is one of the most fruitful sources of brittleness of the horn, which often results in "sand-crack."

The operation of paring out the horse's foot is a matter requiring both skill and judgment, and is, moreover, a work of some labor, when properly performed. It will be found that the operator errs much oftener by removing *too little* than too much; at least it is so with the parts which ought to be removed, which are almost as hard and unyielding as flint, and in their most favorable state, require considerable exertion to cut through.

No general rule can be given applicable to the paring out of the feet of all horses, or even of the feet of the same horse at all times. It would be evidently unwise, for example, to pare the sole as thin in a hot, dry, season, when the roads are broken up, and strewed with loose stones, as would be proper in a moderately wet one, when the roads are well bound and even; for, in the case first named, the sole is in constant danger of being bruised by violent contact with loose stones, and therefore, needs a thicker layer of horn for its protection; while the latter case offers the most favorable surface that the greater part of our horses ever have to travel upon, advantage of which should be taken for a thorough paring out of the sole, in order that the internal parts of the foot may derive the full benefit accruing from an elastic and descending sole; a condition of things very essential to the due performance of their separate functions. To take another illustration: horn grows very freely, especially toward the toe in horses with upright feet and high heels; and such are always benefited by having the toe shortened, the heels lowered, and the sole well pared out; whereas in horses with flat feet and low heels, horn grows sparingly, and the toe of such feet being always weak, admits of very little shortening. Such heels being already too low, they should scarcely be touched with the rasp; and the sole presents such a small quantity of dead horn, that the knife should be used with great discretion.

The corners formed by the junction of the crust and bars should be well pared out, particularly on the inside; for this is the common seat of corn, and any accumulation of horn in this situation must increase the risk of bruising the sensitive sole between the inner part or heel of the coffin bone and the horny sole. Little, if anything, is gained by allowing the bars to project beyond the surface of the sole; the power of resisting contraction cannot possibly be increased by this arrangement, and the projecting rim is left exposed to the danger of being broken and bruised by contact with stones and other hard substances; and the method is further attended with the disadvantages of making the cleaning out of these corners a work of considerable ingenuity with so unwieldly an instrument as a common drawing-knife. It is much preferable to pare them down to a level with the sole, or very nearly so; avoiding, however, every approach to what is styled 'opening out the heels," a most reprehensible practice, which means cutting away the sides of the bars, so as to show an apparent increase of width between the heels, which may for the time deceive the eye, but is in reality a mere deception, purchased at the expense of impaired powers of resistance in the bars and ultimate contraction of the feet. It is palpable that the removal of any portion from the sides of the bars must diminish their substance, and render them weaker, and consequently less able to resist contraction.

The frog should never be cut or pared, except in very rare cases of horses with unusually fast-growing frogs. The first stroke of the knife removes the thin horny covering altogether, and lays bare an under surface, totally unfitted, from its moist, soft texture, for exposure either to the hard ground or the action of the air, in consequence of which exposure it soon becomes dry and shrinks; then follow cracks, the edge of which turning outward forms rags; these rags are removed by the smith at the next shoeing, by which means another similar surface is exposed, and another foundation laid for other rags; and this process continues until finally the protruding, plump, elastic cushion, interposed by nature between the navicular joint and the ground, and so essential to its preservation from injury, is converted by this senseless interference into the dry, shrunk, unyielding apology for a frog, to be seen in the foot of almost every horse that has been regularly shod for a few years. The frog is provided within itself with two very efficient modes of throwing off any superfluous horn with which it may be troubled, and it is very unwise in man to interfere with them. The first and most common of these modes is the separation from the surface of the frog of small, bran-like scales, which becoming dry, fall off in a kind of whitish scurf, not unlike the dust that adheres to Turkey figs; the other, which is upon a large scale, and of rarer occurrence, is sometimes called "casting the frog." A thick layer of frog separates itself in a body, and shells off as deep as a common paring with a knife; but this very important difference is to be noted between the two operations—that nature never removes the horny covering until she has provided another horny covering beneath, so that although a large portion of the frog may have been removed, there still remains behind a perfect frog, smaller, it is true, but covered with horn, and in every way fitted to sustain exposure; while the knife, on the contrary, removes the horny covering, but is unable to substitute any other in its stead. The frog should, therefore, be left to itself; nature will remove the superfluous horn, and the rags do no harm, since, if they are unmolested, they will soon wholly disappear.

The shoe should possess these general features: first, it should be, for ordinary work, rather heavy, in order that it may not be bent by contact with hard, uneven roads; second, it should be wide in the web, and of equal thickness and width from the toe to the heel, that it may as much as possible protect the sole, without altering the natural position of the foot; third, it should be well drawn in at the heels, that it may rest on the bars, and extend to the outer edge of the crust on the outside, and reach beyond the bar nearly to the frog, so that there may be no danger of its pressing on the "corn-place," or angles between the bar and the crust; and fourth, it should in no part extend beyond the outer edge of the crust, lest it strike against the opposite leg when the horse is traveling, or be stepped on by another horse, or be drawn off by a heavy soil.

Such a shoe, and its position on the foot, is shown in the cut opposite.

The shoe should be made as nearly of this form as the shape of the foot will allow; but it is always to be borne in mind, that the shoe is intended for the foot, and not the foot for the shoe, and that it is therefore peculiarly proper to make the shoe to fit the natural form of the foot, instead, as is too often the case, of paring, burning, and rasping the foot until it fits the shoe, which is made according to the smith's notion of what the form of the horse's foot should be. No amount of paring can bring the foot of a horse to an unnatural figure, and also leave it sound and safe for work and use.

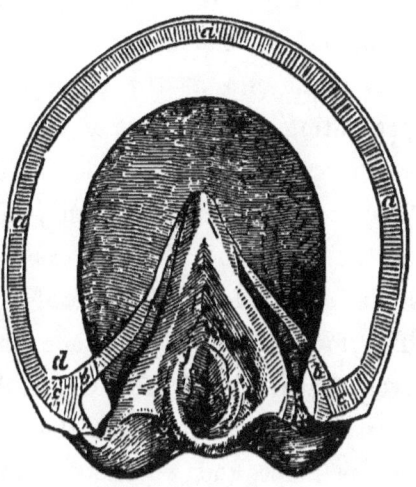

THE POSITION OF THE SHOE.

This cut represents the foot with the shoe rendered transparent, showing what parts of the foot are protected and covered by bringing in the heels of the shoes. *a, a, a*, the crust, with the shoe closely fitted all around. *b, b*, the bars, protected by the shoe. *c, c*, the heels, supported by the shoe. *d*, the situation of corns protected from injury.

The truth really is, that the shape of the shoe cannot by possibility influence the shape of the foot; for the foot being elastic, it expands to the weight of the horse in precisely the same degree, whether it is resting upon the most open or the most contracted shoe. It is the situation of the nails, and not the shape of the shoe, that determines the form of the foot. If the nails be placed in the outside quarter and toe, leaving the heels and quarters on the inside, which are the most expansive portions, free, no shape which we can give to the shoe can of itself change the form of the foot. It must not, however, be inferred from this, that the shape of the shoe is **therefore of no importance**; quite the contrary being the case, as has been already shown. As the shape of the foot is in no degree changed by the form of shoe, that form should manifestly be adopted which produces the greatest number of advantages with the fewest disadvantages.

A small clip at the point of the toe is very desirable, as preventing displacement of the shoe backwards; it need not be driven up hard, as it is simply required as a check or stay. The shoe should be sufficiently long to fully support the angles at the heels, and not so short, as is too often the case, that a little wear imbeds the edge of it in the horn at these parts. The foot surface of the shoe should always have a good flat even space left all around for the crust to bear upon; for it must be remembered, that the crust sustains the whole weight of the horse, and should therefore have a perfectly even bearing everywhere around the shoe. In this space the nail-holes should be punched; and not, as is too generally the case, partly in it, and partly in the seating. In what is technically called "back-holing the shoe," which means completing the openings of nail-holes on the foot surface, great care should be taken to give them an outward direction, so as to allow the points of the nails to be brought out low down in the crust. The remainder of the foot surface should be carefully seated out particularly around the elevated toe, where it might otherwise press inconveniently upon the sole, and the seating should be carried on fairly to the point where the crust and bars meet, in order that there may be no pressure in the seat of corns; the chance of pressure in this situation will be still further diminished by beveling off the inner edge of the heels with a rasp.

The ground surface should be perfectly flat, with a groove

running round the outer edge, just under the plain surface, upon which the crust bears. The principal use of this groove is to receive the heads of the nails that secure the shoe, and prevent their bending or breaking off; it is further useful in increasing the hold of the shoe upon the ground, and should be carried back to the heels.

In fitting the shoe on the foot, it should never while red-hot be burned into its place, as this would so heat the sensitive sole as to produce a serious derangement of its parts; but it may with safety be touched lightly to the foot, that by a slight burning it may indicate those parts where the foot needs paring; indeed, it is necessary to pursue this course in order to make the shoe so exactly fit the foot that there will be no danger of its moving sufficiently to loosen the hold of the nails. The shoe should be made with steel in front, this being sloped backwards to a line running at right angles with the upper slope of the hoof. Old shoes being always worn to about this form, new ones should be so made, and the steel will prevent their being unduly worn.

The shoe having been so fitted that the foot exactly touches it in every part, the next step is to nail it fast to the hoof. Upon the number and situation of the nails which secure it depends the amount of disturbance that the natural functions of the foot are destined to sustain from the shoe. If the nails are numerous, and placed back in the quarters and heels, no form of shoe, however perfect, can save the foot from contraction and navicular disease. If, on the contrary, they are few, and placed in the outside quarter and toe, leaving the inside quarter and heels free to expand, no form of shoe is so bad that it can, from defective form alone, produce contraction of the foot.

Various experiments, which have been made for the purpose of ascertaining how few nails are absolutely necessary under ordinary circumstances for retaining a shoe securely in its place, have satisfactorily established that five nails are amply sufficient for the fore-shoes and seven for the hind. The nails should not be driven high up in the crust, but brought out as soon as possible; they should also be very lightly driven up before the clinchers are turned down, and not, as is generally the case, forced up with all the power which the smith can bring to bear upon them with his hammer. The clinches should not be rasped away too fine, but turned down broad and firm. The practice of rasping the whole surface of the hoof after the clinches have been turned down, should never be allowed; it destroys the covering provided by nature as a protection against the too rapid evaporation of the moisture of the hoof, and causes the horn to become dry and brittle.

The fear, very commonly entertained, that a shoe will be cast almost at every step, unless it is held to the foot by eight or nine nails driven high up into the crust, is utterly groundless, as both theory and practice concur in asserting. If the presence of a nail in the crust were a matter of no moment, and two or three more than are necessary were merely useless, no great reason would exist for condemning the common practice of using too many nails; but it is far otherwise;—the nails separate the fibres of the horn, which never by any chance become united again, but continue apart and unclosed, until by degrees they grow down with the rest of the hoof, and are finally, after repeated shoeings, removed by the knife.

If the clinches chance to rise, they must be at once replaced, as such rising imparts to the nails a freedom of motion which is certain to enlarge the size of the holes; and this mischief is often increased by the violent wrenching from side to side which the shoe undergoes in the process of removal by the smith. As these holes cannot possibly grow down and be removed under three shoeings, it will be found that even with seven nails the crust must always have twenty-one of these separations existing in it at the same time; and as they are often from various causes extended into each other, they necessarily keep it in a brittle, unhealthy state, and materially interfere with the security of the future nail-hold.

By the mode of fastening above advocated the struggle between the expansion of the foot and the resistance of the shoe is entirely overcome; the outer side of the foot, being the only part nailed to the shoe, carries the whole shoe with it at every expansion; while the inner side, being unattached, expands independently of it, whereby all strain upon the nails is avoided, and the foot is left, with respect to its power of expansion, as nearly as possible in a state of nature.

The position of the hind foot and the nature of its office render it less liable to injury than the fore foot, and consequently it less frequently lames. As, however, disease of the navicular bone of this foot is by no means impossible, care should be taken to guard against its contraction by interfering as little as possible with the expansive power of the foot; and this is best done by keeping the nails on the inside as far removed from the heel as convenient, placing four nails in the outer and three in the inner side of the shoe. The holes in the inner side should be punched closer together, and kept more towards the toe than those on the outside, which should be more spread out, as affording greater security of hold to the foot. The shoe should be carefully fitted to the hoof all round, particularly at the heels, which are too often left without any support whatever; and the mischievous custom of turning down the outer heel only must be avoided, because it throws the weight entirely upon the inner quarter, which is the part least able to bear it, and causes much uncomfortable strain to the fetlock joint above. Calkins, even though they are turned down of perfectly even length on each side, (which, however, is rarely done,) are objectionable appendages, and had better be dispensed with, except, perhaps, for very heavy draft, where their ends by entering the ground may prevent the foot from slipping backwards, and may thus enable the toe to obtain a firmer hold.

The form of shoe here referred to, and the position of the nail-holes are shown in the cut annexed.

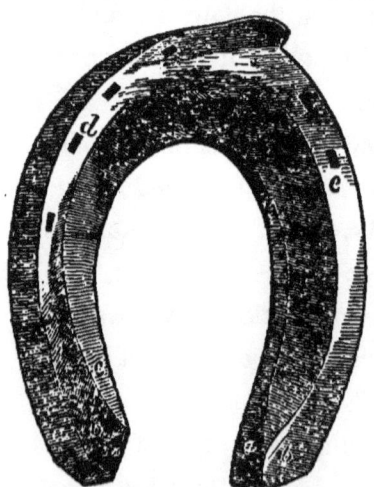

THE PROPER FORM OF A SHOE.

Before leaving this subject it should be remarked, that contracted feet—that is, feet that have shrunken and become narrow at the heels, and of which the frog has become materially reduced in size,—are often, and doubtless most frequently, caused by inflammation arising from improper shoeing. It is the custom of many smiths to "set the shoes well off at the heels;" and to

a. a The heels of an even thickness with the rest of the shoe. *b. b.* Show the points at which the heels of the hoof terminated. *c. c.* The seating carried back, so as to clear the angles at the heels, and leave the seat of corns free from pressure. *d.* The nail-holes placed in the flat surface which supports the crust, where they should always be. *e.* The hindermost nail of the inner side at the inner toe, whereby the whole of the quarter and heel are left free to expand.

carry the seating or level of the upper side of the shoes so far back that the heels, instead of resting on a flat surface, as they would on a properly fitted shoe, rest on the slopes of the seating, which are in this respect simply two inclined planes, so placed that, at each step taken by the horse, his heels must be pressed together, until a greater or less contraction is made manifest, but at too late a period to enable us to remedy the evil; for there is no means by which this contraction of the foot can be cured—although, when it exists only to a slight extent, the internal portions of the foot will sometimes accommodate themselves to its new form. So far as disease is the result of bad shoeing, it can be obviated by so forming the shoe that it will afford a sufficient and perfectly secure and level support for the heels.

ADMINISTERING MEDICINE.

The most common form in which medicine is given to the horse is by means of the BALL, which is an oblong mass of rather soft consistence, yet tough enough to retain its shape, and wrapped up in thin paper for that purpose. The usual weight of the ball is from half an ounce to an ounce, but they may be given of a larger size, if they are made longer but not wider. Every person in charge of horses should know how to give a ball, which is managed either with or without a balling-iron, an instrument seldom wanted, and which sometimes occasions considerable injury to the roof of the horse's mouth. Occasionally, a horse cannot be managed by any other means; but, generally speaking, these instruments only furnish an excuse for bad management. In giving a ball in the ordinary way, the horse's tongue is drawn out of his mouth on the off or right side, and held there firmly with the left hand grasping it as near the root as possible, but to a certain extent yielding to the movement of the horse's head, so as not absolutely to tear it out. While the tongue is thus held, the ball is placed between the fingers and thumb of the right hand, extended in a wedge-like or conical form, so as to pass as far down the swallow as possible; and the hand in this form, with the arm bared to the shoulder, is carried over the root of the tongue till it feels the impediment caused by the contraction of the swallow, when the fingers leave the ball there, and the hand is withdrawn quickly yet smoothly, while at the same moment the tongue is released, and the head is held up till the ball is seen to pass down the gullet on the left side of the neck, after which the head may be released. When the balling-iron is used, the oval ring of which it is composed is passed into the mouth, so as to keep it open, being first well guarded with tow or cloths wrapped around it; the handle is then held in the left hand, together with the halter, so as to steady the head, and yet to keep the horse from biting; and while thus held the hand can be freely carried over the tongue, and the ball be deposited in the pharynx. When a horse is very determined, it is sometimes necessary to keep the iron in the mouth by means of the check-pieces of an ordinary bridle buckled to the sides of the oval ring; but this expedient is seldom required if the halter is firmly grasped with the handle of the iron.

In the usual way the horse to be balled is turned around in his stall, which prevents his backing away from the person in charge; and if the latter is not tall enough, he may stand upon a sound stable-bucket, turned upside down. Balls should be recently made, as they soon spoil by keeping; not only losing their strength, but also becoming so hard as to be almost insoluble in the stomach, and frequently passing through the bowels nearly as they went into the mouth. When hard they are also liable to stick in the horse's gullet. If ammonia, or any other strong stimulant, is given in this way, the horse should not have his stomach quite empty, but should have a little gruel or water just previously; for if this is put off till afterward, the nauseous taste of the ball almost always prevents his drinking. When arsenic forms an ingredient of the ball, it should be given soon after a feed of corn; or a quart or two of gruel should be given instead, just before the ball.

The administration of a DRENCH is a much more troublesome affair than the giving of a ball; and in almost all cases more or less of the dose is wasted. Sometimes, however, a liquid medicine is to be preferred, as in colic or gripes, when the urgent nature of the symptoms demands a rapidly acting remedy, which a ball is not, as it requires time to dissolve; and, besides this, a ball cannot contain any of the spirituous cordials. The best instrument for giving a drench is the horn of the ox, cut obliquely, so as to form a spout. Bottles are sometimes used in an emergency, but their fragile nature always renders them dangerous. In giving a drench, the tongue is held in the same way as for the delivery of a ball, but the head must be more elevated; the drench is then carefully poured into the throat, after which the tongue is let go, but the head still kept up till it is all swallowed. Allowance should always be made for some waste in giving a drench.

In managing horses while in PHYSIC, the horse should in all cases, if possible, be prepared by bran mashes, given for two or three nights, so as to make the bowels rather loose than otherwise, and thus allow the dose to act without undue forcing of the impacted *fœces* backward. If physic is given without this softening process, the stomach and bowels pour out a large secretion of fluid, which is forced back upon the rectum, and met by a solid obstacle which it takes a long time to overcome, and during that interval the irritating purge is acting upon the lining membrane, and often produces excessive inflammation of it. Purging physic should generally be given in the middle of the day, after which the horse should remain in the stable, and have chilled water as often as he will drink it, with bran mashes. By the next morning he will be in a condition to be walked out for an hour, which will set the bowels acting, if they have not already begun. It is usual to tie up the tail with a tape or string, so as to keep it clean. The horse should be warmly clothed, and if the physic does not act after an hour's walk, he may be gently trotted for a short distance, and then taken home; and if still obstinate, he may be exercised again in the afternoon. As soon as the physic operates pretty freely, the horse is to be taken into his stable, and not stirred out again, under any pretense whatever, for forty-eight hours after it has "set," or, in common language, stopped acting. When the purging has ceased, the mashes may be continued for twenty-four hours, with a little corn added to them, and a quantity of hay. The water, during the whole time, should be in small quantities, and chilled; and the clothing should be rather warmer than usual, taking great care to avoid draughts of cold air. Every horse requires at least a three-day's rest for a dose of physic, in order to avoid risk of mischief.

The mode of giving a CLYSTER is now rendered simple enough, because a pump and tube are expressly made for the purpose; and it is only necessary to pass the greased end of the tube carefully into the rectum, for about eight or nine inches, and then pump the liquid up until a sufficient quantity is given. From a gallon to six quarts is the average quantity, but in colic a much larger amount is required.

LOTIONS are applied by means of cloth bandages, if used to the legs; or by a piece of cloth tied over the parts, if to any other surface.

FOMENTATIONS are very serviceable to the horse in all recent external inflammations; and it is astonishing what may be done by a careful person, with warm water alone, and a good-sized sponge. Sometimes, by means of an elastic tube and stop-cock, warm water is conducted in a continuous stream over an inflamed part, as in severe wounds, etc., in which this plan is found wonderfully successful in allaying the irritation, which is so likely to occur in the nervous system of the horse. A vessel of warm water is placed above the level of the horse's back, and a small india-rubber tube leads from it to a sponge fixed above the parts, from which the water runs to the ground as fast as it is over-filled. This plan can be very easily carried out by any person of ordinary ingenuity.

RUNNING AWAY

Vices of the Horse

THE many excellent qualities of the horse are accompanied by some defects, which occasionally amount to vices. These may in part be attributed to natural temper; for man himself scarcely presents more peculiarities of temper and disposition than does the horse. The majority of these disagreeable or dangerous habits in the animal now under consideration are without doubt attributable to a faulty education. The instructor was ignorant and brutal, and the animal instructed becomes obstinate and vicious. It is proposed to mention some of the more glaring of these vices, suggesting in connection with each whatever remedies or palliatives experience has suggested.

(198)

RESTIVENESS.

This stands in the front rank of all the vicious qualities of the horse, being at once the most annoying and the most dangerous of all. It is the direct and natural result of bad temper and worse education; and, like all other habits based upon nature and engrained by education, it is inveterate. Whether it develop itself in the form of kicking, rearing, plunging, bolting, or in any way that threatens danger to the rider or horse, it rarely admits of a cure. The animal may, indeed, to a certain extent be subjugated by a determined rider; or he may have his favorites, or form his attachments, and with some particular person be comparatively or perfectly manageable; but others cannot long depend upon him, and even his master is not always sure of him.

BAULKING OR JIBBING.

This species of restiveness is one of the most provoking vices of the horse, and it can be successfully combated only by a man of the most imperturbable temper. The slightest sign of vexation only increases the evil, and makes the animal more and more troublesome each time that he refuses his work. Many a thick-headed, quick-tempered driver flies into a passion, and beats or otherwise abuses his horse, on the least symptom of baulking, until the animal becomes utterly worthless from a confirmation of the habit.

As a rule it may be stated, that horses baulk from nervousness, or unsteadiness of disposition; if not, indeed, from an over-anxiety to perform their work. Nervous, well-bred horses are more susceptible to the influences which induce baulking, than are colder blooded, more indolent ones. A high-mettled horse, when carelessly driven, will start suddenly against his collar, fail to start his load, draw back from the pain which the concussion causes, rush at it again, and again draw back, until it becomes impossible for his driver to steady him in his collar for a dead pull. If to all this be added a smart cut with the whip, and a fiercely spoken word,—with perhaps a blow over the nose, or a stone in the ear,—every fear or vicious feeling of the horse will be summoned into action, and the animal will become entirely unmanageable, requiring to be left for an hour or two in his position before he gets sufficiently calm to be induced to move. There may, occasionally, be a horse which cannot be made to draw steadily by the most careful treatment; but the cases are exceedingly rare in which gentle treatment and firmness—a patient persistence in mild, authoritative command, and judicious coaxing—would not either prevent the formation of the habit, or cure it when formed.

The prevention of baulky habits lies with the driver. If he jump upon his load, gather up his reins carelessly, flourish his whip, or call out wildly to his horse, he will be quite likely to start him forward with a jerk which will be of no avail to move a heavily laden wagon. The horse thus commences to baulk at a heavy load, and after a certain amount of such treatment, will refuse to draw anything except under the most favorable circumstances. Let any person driving a strange horse, with a load that he is not perfectly sure he can start easily, proceed according to the following directions, and he may be certain that, if the animal be not already a "jibber," he will not make him so, and that if he is one he will have the best chance for getting him along without trouble: He should slowly examine the harness and wagon (partly to accustom the horse to his presence,) gather up the reins gently, speaking to the horse to prevent his starting, get quietly into his seat, and then, if possible, get control of the horse's mouth before allowing him to move, so that when he does step off it may be only at a slow walk. If by a forward movement of the hands he can be made to press very gradually against the collar, and if the whole operation is performed in a cool and unexcited manner, there will be little difficulty in bringing him to a dead pull, from which he will recoil only if the load is a serious tax upon his strength.

If the first attempt fail, wait until your horse has become quiet, and until you have recovered from your own vexation, and then try again. It may be necessary to have the assistance of one or two persons, to start the wagon from behind; but they should not push it until the horse is fairly against the collar.

To cure the habit of baulking is not an easy matter, and it is possible only by the kindest treatment. If the horse show fear by his excited manner, or, by looking about him wildly, that he is expecting a blow, you may be sure that he has received hard usage under similar circumstances, and that he must be convinced by caresses and kind words that you will treat him gently. You must recollect that the horse cannot understand your language; and that, while he is confused, he will misinterpret every sign which you may make to him. He has an idea of your superior power; and, in his fear that you will exercise it, as bad drivers have done before, to his injury, he will not at once feel confidence in your kind intentions. He must feel this confidence, whether it take an hour or all day to convey it to him, before you can do anything to cure him of his trick. If you have him harnessed to a light wagon on a smooth road where it will afford but little resistance, you may by repeated trials convince him that it is a simple, easy matter to draw it; and you should continue to exercise him from day to day with the same light load, and afterward increase it gradually, until you have trained him to a quiet manner of starting, or of going up a hill or elsewhere where he has been accustomed to baulk.

By the same gentle treatment you may start a horse or a team that have baulked under the driving of another person. Request the driver and all spectators to go to the side of the road, and then unfasten the check-reins, hang the reins where they will be easily accessible, but so that they may lie loosely upon the horses' backs, caress them, and allow them to look about and convince themselves that no harm is doing. When they have come properly quiet, go to their heads and stand directly in front of the worst jibber of the team, so that his nose may come against your breast if he start. Turn them gently to the right, without allowing them to tighten their traces, and after caressing them a little, draw them in the same way to the left. Presently turn them to the right, and as you do so, bring them slowly against their collars, and let them go.

Sometimes a horse not often accustomed to baulk, betrays a reluctance to move, or a determination not to move. In such cases, the cause, if practicable, should always be ascertained. He may be overtaxed, his withers may be wrung, or he may be insupportably galled or pained by the harness. Those accustomed to horses know what seemingly trivial circumstances occasionally produce this vice. A horse, whose shoulders are raw, or have frequently been so, will not start with a cold collar; but when the collar has acquired the warmth of the parts upon which it presses, he will go without reluctance. Some determined baulkers have been reformed by constantly wearing a false collar, or strip of cloth around the shoulders, so that the coldness of the usual collar should never be felt; and others have been cured by keeping the collar on night and day, for the animal is not able to lie down completely at full length, which the tired horse is always glad to do. When a horse baulks, not at starting, but while doing his work, it has sometimes been found useful to line the collar with cloth instead of leather; the perspiration is readily absorbed, the substance pressing upon the shoulder is softer, and it may be far more accurately eased off at a tender place.

BITING.

This is either the consequence of natural ferocity, or a habit acquired from the foolish and teasing play of grooms and stable-boys. When a horse is tickled and pinched by thoughtless and mischievous youths, he will at first pretend to bite his tormentors; by degrees he will proceed further, and actually bite them, and very soon after that he will then be the first to challenge to the combat, and without provocation will seize the first opportunity to grip the careless teaser. At length, as the love of mischief is a propensity too easily acquired, this war, half playful and half in earnest, becomes habitual to him, and degenerates into absolute viciousness.

It is seldom that any thing can be done in the way of cure. Kindness will aggravate the evil, and no degree of severity will correct it. Biters have been punished until they have trembled in every joint, and were ready to drop, but this treatment scarcely ever cures them. The lash is forgotten in an hour, and the horse is as ready and determined to repeat the offense as before. He appears unable to resist the temptation, and in its worst form biting is a species of insanity.

Prevention, however, is in the power of every proprietor of horses. While he insists upon gentle and humane treatment, he should systematically forbid this horse-play.

KICKING.

This, as a vice, is another consequence of the culpable habit of teasing the horse. That which is at first simply an indication of annoyance at the pinching and tickling of the groom, and without any design to injure, gradually becomes the expression of anger, and the effort to do mischief. The horse, also, too soon recognizes the least appearance of timidity, and takes advantage of the discovery.

PARTICULARLY DANGEROUS.

Some horses acquire, from mere irritability and fidgetiness, a habit of kicking at the stall or the bail, and particularly at night. The neighboring horses are disturbed, and the kicker gets swelled hocks, or some more serious injury. This is a habit very difficult to correct, if it is allowed to become established. Mares are much more subject to it than horses.

Before the habit is inveterately established, a thorn-bush or a piece of furze fastened against the partition or post will sometimes effect a cure. When the horse finds that he is pretty severely pricked, he will not long continue to punish himself. In confirmed cases it may be necessary to have recourse to the log, but the legs are often not a little bruised by it. A rather long and heavy piece of wood attached to a chain has been buckled above the hock, so as to reach about half-way down the leg. When the horse attempts to kick violently, his leg will receive a severe blow; this, and the repetition of it, may after a time teach him to be quiet.

Kicking in harness is a much more serious vice. From the least annoyance about the rump or quarters, some horses will kick at a most violent rate, and destroy the bottom of the chaise, and endanger the limbs of the driver. Those that are fidgety in the stable are most inclined to do this. If the reins chance to get under the tail, the violence of the kicker will often be most outrageous; and while the animal presses down his tail so tightly that it is almost impossible to extricate the reins, he continues to plunge until he has demolished every thing behind him.

This is a vice standing foremost in point of danger, and one which no treatment will always conquer. It is altogether in vain to attempt coercion. If the shafts are very strong and without flaw, or if they are plated with iron underneath, and a stout kicking-strap resorted to, which will barely allow the horse the proper use of his hind limbs in progression, but not permit him to raise them sufficiently for the purpose of kicking, he may be prevented from doing mischief.

REARING.

This vice is not very common, at least in a dangerous form, and can generally be prevented by the use of the martingale.

In the case of saddle-horses, another good prevention is, when the horse is about to rise, to touch him with the spur on one side only; this will cause him to stop to lift the hind leg on that side, and if he persists in his attempt the spurs may be used vigorously, first on one side, and then on the other, but not so fast as to prevent the horse from raising his hind legs alternately, as he is spurred. The least touch of the curb-bit will cause some vicious and badly trained horses to rear, while those which have been thoroughly trained will rear slightly, to a great height, or not at all, as their rider may desire; but it is obvious that horses so delicately trained should not be ridden by unskillful persons, lest the awkwardness of the rider should cause unexpected curveting.

The remedy of some breakers, that of pulling the horse backward on a soft piece of ground should be practiced by reckless and brutal fellows alone. Many horses have been injured in the spine, and others have broken their necks, by being thus suddenly pulled over; while even the fellow who fears no danger, is not always able to extricate himself from the falling horse. If rearing proceeds from vice, and is unprovoked by the bruising and laceration of the mouth, it fully partakes of the inveteracy which attends the other divisions of restiveness.

PULLING BACK ON THE HALTER.

This is a vice which has probably arisen from the horse having, at some time, broken a weak halter in a fit of impatience. The only safe cure for it, and this is not always successful, is to tie the horse with a very strong halter, which it will be impossible for him to break; finding that his efforts are futile, he will, after a time, generally desist from pulling—though some incorrigible brutes will try every new halter as soon as they are fastened, and will break it if possible.

RUNNING AWAY.

Some headstrong horses will occasionally endeavor to bolt with the best rider; others, with their wonted sagacity, endeavor thus to dislodge only the timid or unskillful one. Some are hard to hold, or bolt only during the excitement of a trial of speed, or the like; others will run away, prompted by vicious propensity alone. There is no certain cure here. The only method which affords any probability of success is, to ride such a horse with a strong curb and sharp bit; to have him always firmly in hand; and if he will run away, and the place will admit of it, to give him (sparing neither curb, whip, nor spur,) a great deal more running than he likes.

VICIOUS TO CLEAN.

It would scarcely be credited to what an extent this exists in some horses that are otherwise perfectly quiet; it is only at great hazard that they can be cleaned at all. The origin of this is probably some maltreatment. There is, however, a great difference in the sensitiveness of the skin in different horses. Some seem as if they could scarcely be made to feel the whip, while others cannot bear a fly to light upon them without an expression of annoyance. In young horses the skin is peculiarly delicate. If they have been curried with a broken comb, or hardly rubbed with an uneven brush, the recollection of the torture they have felt makes them impatient and even vicious during every succeeding operation of the kind. Many grooms, likewise, seem to delight in producing these exhibitions of uneasiness and vice, although, when they are carried a little too far, and at the hazard of the limbs of the groom, the animals that have been almost tortured into these manifestations of irritation, are brutally kicked and punished.

This, however is a vice that may be conquered. If the horse is dressed with a lighter hand, and wiped rather than brushed, and the places where the skin is most sensitive are avoided as much as thorough cleanliness will allow, he will gradually lose the recollection of former ill-treatment, and become tractable and quiet.

In those instances where the skin is so irritable that the horse really endures a great deal of misery every time he is cleaned besides requiring needlessly the expenditure of a great amount of muscular exertion, the remedy is very simple; instead of being curry-combed and wiped, the horse should be merely washed over with warm water on his coming in warm from a journey, then gently scraped and covered with a rug. The warmth of the body will very soon dry the skin.

VICIOUS TO SHOE.

The correction of this is more peculiarly the business of the smith; yet the master should diligently concern himself with it, for it is more often the consequence of injudicious or bad usage, than of natural vice. The vice is certainly a bad one, and it very materially diminishes the value of the horse; for it is a habit which generally gets worse at each time of shoeing. It is not so much the kicking of the horse that is to be feared, but the animal will bear his whole weight on the foot requiring to be shod, so that the smith is unable to lift it up, or afterward to support it; beside which the animal will keep continually kicking or endeavoring to get the foot away, to the imminent danger of the limbs of the unfortunate operative. This deplorable and vicious habit is greatly increased, if not altogether produced, by rough usage at the early shoeings, and it generally gets worse at each time of shoeing, so that the horse is often rendered at last completely worthless.

It may be expected that there will be some difficulty in shoeing a horse for the first few times, as it is an operation that gives him a little uneasiness. The man to whom he is most accustomed should go with him to the forge; and if another and steady horse is shod before him, he may be induced more readily to submit. It cannot be denied that, after the habit of resisting this necessary operation is formed, force may sometimes be required in order to reduce our rebellious servant to obedience; but there can be no manner of question that the large majority of horses vicious to shoe are rendered so by harsh usage, and by the pain of correction being added to the uneasiness of shoeing. It should be a rule in every forge, that no smith should be permitted to strike a horse, much less to twitch or gag him, without the master-farrier's order; and that a young horse should never be twitched or struck. There are few horses that may not gradually be rendered manageable for this purpose by mildness and firmness on the part of the operator; they will soon understand that no harm is meant, and they will not forget their usual habit of obedience; but if the remembrance of corporeal punishment is connected with shoeing, they will always be fidgety, and occasionally dangerous.

CRIB-BITING.

This is a very unpleasant habit, and a considerable defect, although not so serious as it is often represented. The horse lays hold of the manger with his teeth, violently extends his neck, and then, after some convulsive action of the throat, a slight grunting is heard, accompanied by a sucking or drawing in of air. It is not an effort at simple eructation, arising from indigestion; it is the inhalation of air. It is that which takes place with all kinds of diet, and when the stomach is empty as well as when it is full.

The effects of crib-biting are plainly perceptible. The teeth are injured and worn away, and that, in an old horse, to a very serious degree. A considerable quantity of grain is often lost, for the horse will frequently crib with his mouth full of it, and the greater part will fall over the edge of the manger. Much saliva escapes while the manger is thus forcibly held, the loss of which must be of serious detriment in impairing digestion The crib-biting horse is notoriously more subject to colic than other horses, and that of a kind difficult of treatment and peculiarly dangerous. Although many a crib-biter is stout and strong, and capable of all ordinary work, these horses do

MUZZLE FOR A CRIB-BITER.

not generally carry as much flesh as others, and have not their endurance; on these accounts, crib-biting has been, and very properly, decided by the highest authority to be unsoundness.

It is moreover one of those tricks which are exceedingly contagious. Every companion of a crib-biter in the same stable, is likely to acquire the habit, and it is the most inveterate of all habits. The edge of the manger will in vain be lined with iron, or with sheep-skin, or with sheep-skin covered with tar or aloes, or any other unpleasant substance. In spite of the annoyance which these may occasion, the horse will persist in his attack on the manger. A strap buckled tightly round the neck, by compressing the windpipe, is the best means of preventing the possibility of this trick; but the strap must be constantly worn, and its pressure is apt to produce a worse affection, viz., an irritation of the windpipe, which terminates in roaring.

Some have recommended turning out for five or six months; but this has never succeeded, except with a young horse, and then but rarely. The old crib-biter will employ the gate for the same purpose as the edge of his manger, and he will often gallop across a field for the mere purpose of having a bite at the rail. Medicine is altogether thrown away in such a case.

The only remedy is a muzzle, with bars across the bottom; sufficiently wide to enable the animal to pick up his corn and to pull his hay, but not to grasp the edge of his manger. If this is worn for a considerable period, the horse may be tired of attempting that which he cannot accomplish, and for a while forget the habit; but in a majority of cases the desire of crib-biting will return with the power of gratifying it.

The causes of crib-biting are various, and some of them beyond the control of the owner of the horse. It is often the result of imitation; but it is more frequently the consequence of idleness. The high-fed and spirited horse must be in mischief, if he is not usefully employed. Sometimes, but not often, it is produced by partial starvation; and another occasional cause is the frequent custom of dressing the horse, even when the weather is not severe, in the stable,—thus enabling the animal to catch at the edge of the manger, or at that of the partition on each side, if he has been turned.

WIND-SUCKING.

This closely resembles crib-biting, and arises from the same causes; the same purpose is accomplished, and the same results follow. The horse stands with his back bent, his head drawn inward, his lips alternately slightly opened and then closed, and a noise is heard as if he were sucking. It appears quite probable, judging from the same comparative want of condition and the flatulence noted in connection with the last habit, that either some portion of wind enters the stomach, or there is an injurious loss of saliva.

This vice diminishes the value of the animal nearly as much as crib-biting; it is equally as contagious and inveterate. The only remedies—and they will seldom avail—are tying the head up, except when the horse is feeding, or putting on a muzzle with sharp spikes toward the neck, which will prick him whenever he attempts to rein his head in for the purpose of wind-sucking.

OVERREACHING.

This unpleasant noise known also by the name of "clicking," is occasioned by the toe of the hind foot, or the inner edge of the inside of its shoe, striking upon the heel of the coronet of the fore foot. The preventive treatment is the beveling, or rounding off, of the inside rim or edge of the hind shoe. The cure is, the cutting away of the loose parts, the application of Friar's balsam, and protection from the dirt.

Some horses, particularly young ones, overreach so as to strike the toes of the hind shoes against the fore ones, which is termed "clinking." Keeping up the head of the horse does something to prevent this; but the smith may do more by shortening the toe of the hind shoes and having the web broad. When they are too long, they are apt to be torn off; when too narrow, the hind foot may bruise the sole of the fore one, or may be locked fast between the branches of the fore shoe.

NOT LYING DOWN.

It occasionally happens that a horse will seldom or never lie down in the stable. He sometimes continues in apparent good health, and feeds and works well; but generally his legs swell, and he becomes fatigued sooner than another horse. If it is impossible to let him loose in the stable, or to put him into a spare box, nothing can be done to obviate the difficulty. No means, gentle or cruel, will force him to lie down. The secret is, that he is tied up, and either has never dared to lie down through fear of the confinement of the halter, or he has been cast in the night and severely injured. If he can be suffered to range the stable, or have a comfortable box in which he may be loose, he will usually lie down the first night. Some few horses, however, will lie down in a stable, and not in a loose box. A fresh, well-made bed will generally tempt the tired horse to refresh himself with sleep.

It may be observed in this connection, that the basis of support afforded by the four extremities is so considerable in the horse, that he is able to sleep in a standing position, and some horses have even been known to preserve their health, strength, and condition, although they were never known to lie down. At the same time, it is undeniable, that an animal that will quickly lie down and take his rest, as a general rule, preserves his condition, and is better fitted for exertion.

SHYING.

This most dangerous habit is sometimes the effect of fear, and sometimes is a downright vicious propensity; and there are many horses which commence the practice through fear and end by becoming viciously disposed to indulge in it, in consequence of sheer mismanagement. The young colt is almost always more or less shy, especially if he is brought at once from the retired fields in which he was reared to the streets of a busy town.

DISAGREEABLE AND DANGEROUS.

There are, however, numberless varieties of shyers, some being dreadfully alarmed by one kind of object, which to another is not at all formidable. When a horse finds that he gains his object by turning around, he will often repeat the turning without cause, pretending to be alarmed, and looking out for excuses for it. This is not at all uncommon, and with timid riders leads to a discontinuance of the ride, by which the horse gains his end for the time, and repeats the trick upon the first occasion. In genuine shying from fear, the eyes are generally more or less defective; but sometimes this is not the cause, which is founded upon a general irritability of the nervous system. Thus, there are many that never shy at meeting wagons, or other similar objects, but which almost drop with fear on a small bird flying out of a hedge, or any other startling sound. These are also worse, because they give no notice, whereas the ordinary shyer almost always shows by his ears that he is prepared to turn.

For shyers the only remedy is, to take as little notice as possible, to make light of the occurrence, speak encouragingly, yet rather severely, and *to get them by the object in one way or another.* If needful, the *aid* of the spur and whip may be called in, but not as a *punishment.* If the horse can be urged to go by the object at which he is shying without the whip or spur, so much the better; but if not, he must be compelled to do so by their use. Wherever fear is the cause of shying, punishment only adds to that fear; but where vice has supplanted fear, severity should be used to correct it.

As a general rule, the whip need never be used, unless the horse turns absolutely round, and not then unless there is reason to suspect that he is pretending fear. If he will only go by the object, even with "a wide berth," as the sailors say, he may be suffered to go on his way unpunished; and nothing is so bad as the absurd severity which some horsemen exercise after the horse has conquered his reluctance, and passed the object. At this time he should be praised and petted, with all the encouragement which can be given; and on no account should he be taught to make those rushes which are so commonly seen on the road, from the improper use of whip and spur. If punishment is necessary at all, it must be used beforehand; but it often happens that the rider cannot spare his whip-hand until the shying is over; and then, in his passion, he does not reflect that the time has passed for its employment.

Shying on coming out of the stable is a habit that can rarely or never be cured. It proceeds from the remembrance of some ill-usage or hurt which the animal has received in the act of proceeding from the stable, such as striking his head against a low door-way, or entangling the harness.

When the cure, however, is early attempted, it may be so far overcome that it will be unattended with danger or difficulty. The horse should be bridled when led out or in. He should be held short and tight by the head, that he may feel that he has not liberty to make a leap, and this of itself is often sufficient to restrain him. Punishment, or a threat of it, will be highly improper. It is only timid or high-spirited horses that acquire the habit, and rough usage invariably increases their agitation and terror.

PAWING.

Some hot and irritable horses are restless even in the stable, and paw frequently and violently. Their litter is destroyed, the floor of the stable broken up, the shoes worn out, the feet bruised, and the legs sometimes sprained. If this habit does not exist to any great extent, yet the stable never looks well. Shackles are the only remedy, with a chain sufficiently long to enable the horse to shift his posture, or move in his stall; but these must be taken off at night, otherwise the animal will seldom lie down. Unless, however, the horse possesses peculiar value, it will be better to dispose of him at once, than to submit to the danger and inconvenience that he may occasion.

ROLLING.

This is a very pleasant and perfectly safe amusement for a horse at grass, but cannot be indulged in the stable without the chance of his being dangerously entangled with the collar, rein, or halter, and being cast. Yet, although the horse is cast, and bruised, and half strangled, he will roll again on the following night and continue to do so as long as he lives. The only remedy is not a very pleasant one for the horse, nor always quite safe; yet recourse must be had to it, if the habit of rolling is inveterate. The horse should be tied with length enough of halter to lie down, but not to allow of his head resting upon the ground; because, in order to roll over, a horse is obliged to place his head quite down upon the ground.

SLIPPING THE COLLAR OR HALTER.

This is a trick in which many horses are so well accomplished, that scarcely a night passes without their getting loose. It is a very serious habit, for it enables the horse sometimes to gorge himself with food to the imminent danger of producing staggers; or it exposes him, as he wanders about, to be kicked and injured by the other horses, while his restlessness will often keep the whole team awake. If the web of the halter, being first accurately fitted to his neck, is suffered to slip only one way, or a strap is attached to the halter and buckled round the neck, but not sufficiently tight to be of serious inconvenience, the power of slipping the collar will be taken away.

STUMBLING.

That person must either be a skillful practitioner, or a mere pretender, who engages to remedy this habit. If it arise from a heavy forehand, and the fore legs, being too much under the horse, no one can alter the natural frame of the animal; if it proceeds from tenderness of foot, grogginess, or old lameness, these ailments are seldom cured. Also, if it is to be traced to habitual carelessness and idleness, no whipping will rouse the drone. A known stumbler should never be ridden or driven by any one who values his safety or his life. A tight hand or a strong bracing-rein are precautions that should not be neglected, although they are generally of little avail; for the inveterate stumbler will rarely be able to save himself, and this tight rein may sooner and further precipitate the rider. If after stumbling the horse suddenly starts forward, and endeavors to break into a short trot or canter, the rider may be assured that others before him have fruitlessly endeavored to remedy the nuisance.

If the stumbler has the foot kept as short, and the toe pared as close as safety will permit, and the shoe is rounded at the toe, or has that shape given to it which it naturally acquires in a fortnight from the peculiar action of such a horse, the animal may not stumble quite so much; or if the disease which produced the habit can be alleviated, some trifling good may be done; but in almost every case the stumbler should be got rid of, or put to slow and heavy work. If the latter alternative is adopted, he may stumble as much as he pleases, for the weight of the load and the motion of the other horses will keep him upon his legs.

UNSTEADINESS WHILE BEING MOUNTED.

When this merely amounts to eagerness to start—very unpleasant, indeed, at times, for many a rider has been thrown from his seat before he was fairly fixed in it—it may be remedied by an active and good horseman. It oftentimes happens that while the elderly, inactive, and fearful man is engaged in making more than one ineffectual attempt to vault into the saddle, the horse is dancing about to his annoyance and danger; but no sooner is the animal transferred to the management of a younger and more agile rider, than he becomes perfectly subdued. Severity will here, more decidedly than in any other case, do harm. The rider should be fearless; he should carelessly and confidentially approach the horse, mount at the first effort, and then restrain him for a while; patting him, and not allowing him to proceed until he becomes perfectly quiet. Horses of this kind should not be too highly fed, and should have sufficient daily exercise.

When the difficulty of mounting arises, not from eagerness to start, but from unwillingness to be ridden, the sooner that horse is disposed of the better. He may be conquered by a skillful and determined horseman; but even he will not succeed without frequent and dangerous contests that will mar all the pleasures of the ride.

THE SICK HORSE

DISEASES and their REMEDIES

UNDER this head it is proposed to treat of the various diseases which horse-flesh is heir to, together with their symptoms, and to offer such remedies as personal experience, or the authority of others in whom implicit confidence may be placed, suggests as the most efficacious. For convenience of discussion, these diseases are arranged in the present work under the heads of diseases of the mouth; of the respiratory organs; of the stomach and intestines; of the limbs; of the urinary organs; of the feet and legs; of the heart; of the head; and of the eye;—placing under the head of miscellaneous such as do not appropriately fall under either of the foregoing diseases.

DISEASES OF THE MOUTH.

LAMPAS.

This term is used to designate a fullness or swelling of the bars or roof of the mouth, caused by the cutting of the teeth. Lampas will be found in all colts, although in many the slight inconvenience occasioned by it attracts little or no attention. In others, however, the great tenderness of the parts affected causes the animal to refuse his food, in consequence of which he is by many compelled to submit to an operation equally cruel and unnecessary—that is, no less than burning out the bars of the mouth with a red-hot iron, thereby destroying the functions of the part, and leaving the mouth sore for some time afterward. This mode of treatment has been practised for years, and is even at the present day almost the only one in vogue, although it is of no practical benefit whatever, but, on the contrary, is often very injurious. In the case of the child similarly affected, the humane practitioner seldom does more than to lance the gums. This, certainly, is a more rational mode of operating, and the author's experience convinces him that if the parts inflamed in the case of the horse be simply lanced, the swelling will soon subside, and the horse partake of his food as usual. A common pocket-knife will answer the purpose quite well; and after the lancing the mouth should be washed with a solution of the tincture of myrrh, two ounces to a pint of water, or a solution of alum in water. This should be repeated twice a day for three or four days, during which time give bran mashes or flax-seed gruel, and, if procurable, a small quantity of new grass. No hay, corn, or oats, should be given for a week; at the expiration of which period the teeth will be in a condition to masticate such food.

INFLAMED GUMS.

Occasionally the gums of very young horses, when cutting their teeth, become exceedingly tender, sore, and swollen. As this is principally confined to the yearling, it is generally overlooked by the owner. The treatment in such cases is to cut the gum through to the tooth immediately under it with a lancet or common pocket-knife. The gum being thus broken, the tooth comes through with little pain.

BAGS OR WASHES.

These are soft, puffy swellings of the membrane of the mouth, lining the lips, just within the corners of the mouth. This disease is generally caused by the bearing rein being too tight. They are cured by cutting off a portion of the swelling with a pair of scissors or a knife; after which the parts should be dressed with a little salt, or powdered alum. This generally proves successful.

ULCERS IN THE MOUTH.

Horses, during the process of breaking, are frequently hurt by the pressure of the bit upon the under jaw a little in front of the first molar tooth; in consequence of which the periosteum, or thin fibrous membrane covering the bone, often becomes involved in the inflammation, the bone itself not always escaping injury, a neglect of which occasionally causes the bone of the jaw to become carious or decayed; sinuses, or pipe-like openings, are sometimes formed, which becoming filled with masticated food, become fetid and often occasion troublesome sores. Grooms on discovering this sore, generally attribute it to what is commonly called squirrel grass, or wild barley. If the sore is confined to the gum alone, it should be washed frequently, and dressed with a little tincture of myrrh; but when the bone is affected, it must be examined carefully with a probe, and if found rough, or presenting small openings, the bone must be exposed, and all the diseased parts removed, after which the tincture of myrrh should be used for a dressing. Such operations should be performed by a qualified veterinary surgeon, if one is to be had; otherwise more injury may be done by the bungling operator than would be occasioned by the disease. If such services cannot be procured, caustic silver, or lunar caustic, should be applied to the diseased bone. If the caustic is not readily obtainable, the red-hot iron will answer the purpose as well, or even better. Butter of antimony, placed on a little cotton or tow, and packed in the sore, is an excellent application, as it hastens a separation or exfoliation of the diseased bone, thus enabling the parts soon to heal.

SORE MOUTH.

This is often caused by the bit's cutting or bruising the lips at the angles of the mouth. In carelessly balling horses, also, the under part of the tongue sometimes becomes injured, which frequently escapes notice until the animal refuses his food, and the tongue becomes tender and swollen. In such cases, wash the mouth clean, and sprinkle a teaspoonful of table salt on the sore; the tincture of myrrh occasionally applied will hasten the cure.

CUT TONGUE.

The tongue sometimes becomes bruised from the sudden jerking of the lines in the hands of a careless or obstinate driver, or it may happen from tight reining; that portion of the tongue upon which the bit rests becoming bruised and ulcerated, and the frequent use of the bit keeping up the irritation, until the tongue, in some cases, becomes almost separated by ulceration before it is discovered. Alum water, saltpetre, and tincture of myrrh are the proper dressings.

UNEVEN TEETH.

The molar teeth frequently become very uneven upon their faces or grinding surfaces, in consequence of the *crusta petrosa* wearing away too rapidly and often leaving deep cavities in the teeth, which become filled with food and soon prove a great source of annoyance by interfering with proper mastication. This occurs more particularly in old horses. The upper molar teeth being well protected on the outer surface with enamel, wear less rapidly than the lower ones which are protected upon the inner side. In consequence of this the upper teeth often become very sharp upon the outside, and when the reins are drawn up the cheeks are forced upon these sharp edges and become sore and often lacerated, while the lower ones becoming sharp on the inside edges, lacerate the tongue in a similar manner. The horse from this cause often refuses his food, since mastication causes him severe pain. He soon begins to lose flesh, the digestive organs become deranged, the skin becomes tight, and the animal is perhaps doctored for bots, worms, and the like.

In all these cases the tooth-rasp becomes necessary, which is an instrument made concave, or hollow, upon one side, and convex, or rounding, on the other, with a long handle attached. The rasp is upon the hollow side, the round side and the edges being perfectly smooth so as not to wound the cheeks or tongue when used. With this instrument the sharp corners of the teeth are easily taken off, and the horse is enabled to feed again in the proper manner. If the teeth are in this condition, no medicine is of any avail; all the condition powders in the world will not benefit in the slightest degree; the tooth-rasp is the only remedy that will prove serviceable.

QUIDDING.

This disease, if disease it may be called, is generally caused by the irregular wear of the teeth already mentioned; or it may arise from caries of the teeth, or from a diseased state of the muscles of deglutition. "I have seen," says White, "at the kennel the jaw of a horse which died literally from starvation in consequence of a disease of the grinding teeth, which appeared to have been brought on by feeding on coarse woody hay, containing the stocks of thistles, docks, &c. This animal was what dealers term a quidder, for the muscles of deglutition were at least so affected that he was incapable of swallowing; and after fruitless attempts to chew his food it was thrown out into the manger in a ball or quid, and a great deal of imperfectly chewed hay had been forced into the cavities formed at the roots of some of the grinding teeth." The tooth-rasp sometimes proves a perfect cure in such cases.

WOLF TEETH.

Very erroneous opinions are entertained by horsemen, and even by veterinary surgeons, in reference to these teeth, and various theories have from time to time been set afloat regarding them, arising, for the most part, from a want of proper investigation.

These teeth are natural to all horses, and make their appearance between the first and fifth year. They are not supernumerary teeth, as has been stated by some writers, but are natural teeth found in all colts. The germs of these teeth will be found in the foal at birth, and developed in the jaw of the yearling ready to make their way through the gums. In an examination of at least one hundred heads of colts that have died under eighteen months of age, the author has found in every instance either natural wolf teeth, or the germs from which they are developed. It is a mistaken idea, that these teeth exert any influence over the eyes. Nature never placed them in their position for the purpose of injury. In cases where the eye is supposed to be affected by them, it is simply necessary to treat the eye for inflammation, and allow the teeth to remain. As a general rule they do not remain in the jaw long after being cut; having performed their function, whatever it may be, they fall out and are therefore seldom found. Their removal can do no harm but it is an entirely unnecessary operation.

CARIES OF THE TEETH

The teeth of horses, as has already been stated, are made up of three substances, the enamel, the bone, and the crusta petrosa; and in consequence of their peculiar arrangement and the inability of the animal to inform us of his sufferings, this disease frequently becomes much more serious than in man. Its operation, besides, is quite different upon the teeth of horses from what it is upon the human teeth. In the human subject caries is found, in a large majority of cases, making its appearance as a dark spot between the teeth, on one side of the crown, gradually working inwards, destroying the bone in its progress, and leaving the enamel a mere shell upon the outside of the tooth, while the roots generally remain in a comparatively sound condition during the progress of decay. In the horse, however, caries is a very different thing, as far as its effects are concerned. It makes its appearance upon some one or more of the indentations or depressions upon the face of the tooth, attacking the crusta petrosa, (a substance not found in the human tooth,) and extending from the face through the entire length of the tooth, splitting it up into several thin plates, in consequence of which abscesses often form at the roots of such teeth, which, being prevented from dis-

THE BLOODED MARE FASHION AND FOAL.

charging into the mouth by the food that fills up the cavity, generally find an opening into the nose, discharging their fetid matter through that channel. The animal while in this condition is often treated for catarrh, commonly called distemper. The discharge still continuing, and becoming more and more fetid, the animal is at last supposed to be in a glandered condition and killed.

The first case of this kind which came under the author's notice occurred in the year 1853. Having occasion to visit the yard where dead animals are boiled, the peculiar appearance of one horse lying upon the ground attracted his attention. Upon inquiry he learned that he had been killed as a glandered horse; but failing to recognize any such marks as might be expected in that disease, he made a very careful examination of the head and found the real cause of trouble to be, not glanders, but a carious tooth, of which but three small ribbon-like fragments remained. This horse was but seven years old. An abscess had formed at the root of the tooth, discharging itself into the nostril, whence it was ejected. Another horse, with similar symptoms, pronounced glandered by two eminent veterinary surgeons, was destroyed at the same place in the year 1859. The author's examination disclosed the fact, that the first two molar teeth were almost entirely destroyed by caries, and that a large abscess had formed at their roots, which extended into and completely closed up one nostril, causing an immense tumor on the right side of the head.

The difficulty of examining the molar teeth of the horse, together with the silence of veterinary authors on this important subject, are the only assignable reasons for the little information given us relative to a disease of such common occurrence.

Indeed, the author has frequently been called upon to treat horses laboring under this disease, without a suspicion ever being entertained of its true nature.

A case of this kind came under his notice in the winter of 1858, while on a visit to Jackson, Michigan. He was called to see a bay mare kept for livery purposes, having a discharge from the right side of the face some two inches below the eye, which had existed for about two years. The discharge was of so fetid a character that the animal was rendered unfit for use, and she was consequently turned upon the common to die or get well, as the chances might be, all known modes of treatment having been previously adopted without any beneficial results. He discovered, upon examination, a carious tooth, which was removed, and in a short time the animal became well. During the winter of 1859, a brown mare, belonging to a gentleman in Germantown, Pennsylvania, was sent to the Clinic of the Philadelphia Veterinary College, having been pronounced glandered by a veterinary surgeon and ordered to be killed. Upon examination a large abscess was discovered opening into the nose, together with two carious teeth—the first and second molars of the right side. The mare was cast, and ten pieces of carious teeth removed; the cavity was then well cleaned out, and tow saturated with tincture of myrrh filled in, removing and cleaning every day. Some four weeks subsequently, the animal was sold for one hundred and fifty dollars, sound as a bell; though previously to this operation she could not have been sold at any price. Many similar cases could be mentioned, but the foregoing will serve to show the necessity of making a thorough examination of an animal before pronouncing sentence of death upon it.

Acidity of the fluids of the mouth is generally—and, as the author believes, correctly—assigned as the cause of caries of the teeth. The symptoms are fetid discharges from the nose, obstructed respiration, improper mastication of the food, passing the oats or corn whole, quidding, drowsiness, loss of flesh, staring coat, hide-bound, tossing to and fro of the head, stopping short on the road, starting suddenly, and at times becoming almost frantic. All these symptoms, however, must not be expected to be found in the same case, as different horses are differently affected by the disease. One is drowsy, feeds daintily at times, and again ravenously; another is at times wild, so as to be almost unmanageable. Many of these symptoms occur in other diseases besides those of the teeth; but their presence suggests the necessity for an examination of the mouth, and particularly of the molar teeth, which may be done by passing the hand along the upper molar teeth inside of the cheek, thus enabling the examiner to detect the presence of caries without difficulty.

EXTRACTING TEETH.

When a carious tooth, or one so unequally worn as to cause mischief, is discovered, its removal is necessary to the restoration of the animal's health. In order to accomplish this, the horse must be cast, and the age of the animal considered, in order to make choice of proper instruments. If he is young, say from four to six years, an instrument made similar to the key used by surgeon dentists, is the best adapted; if he is old, a pair of forceps of large size, made in the same manner as the tooth-forceps of dentists, will answer, as the roots of the teeth in old horses are comparatively short, and therefore may be easily extracted.

DISEASES OF THE RESPIRATORY ORGANS

INFLAMMATION.

The diseases of the respiratory organs and air passages are generally of an inflammatory type. In order to fully understand the various diseases to which these important organs are subject, a few remarks regarding the nature of inflammation, its progress, &c., may not be out of place in a work like the present.

Inflammation, then, is a state of altered nutrition, an increased vascularity and sensibility of the parts involved, together with a tendency to change of structure. The symptoms are swelling, pain, heat, and redness where the parts are not covered with hair. The redness is in consequence of a redundancy of blood in the inflamed part, which distends the small capillaries with red particles of blood. When the inflammation is acute, the parts present a bright red or crimson hue; when it is chronic, they are of a dark or purplish red color. As the various terms employed by authors to indicate the various degrees are uninteresting to the general reader, no attempt at detail is here made.

The sensation of pain is mainly due to a stretching of the nerves by the distended blood-vessels. It differs in its character and intensity according to the parts involved, varying from a burning, throbbing, sharp, and lacerating pain to a mere sense of heat, soreness, and a dull sensation of pain. The heat in inflammation is supposed to arise from an increased quantity of blood in the inflamed part. The swelling in the early stage is due to the increased quantity of blood, and afterward to the effusion which takes place in all loose tissues. By inflammation all the various structures of the animal economy may be so altered as to interfere with the performance of their natural functions; in some cases by a permanent thickening of the parts, and in others by adhesion and the like.

By the aid of auscultation, that is, the application of the ear to the parts to be examined, the slightest change in the normal and healthy condition of the respiratory organs may be detected, and the various parts involved in inflammatory action may be pointed out with a considerable degree of certainty. With thus much of introduction we proceed to the consideration of the various diseases naturally falling under our present division.

SORE THROAT.

Sore throat is a common attendant upon catarrhal affections. When it is confined to that portion of the throat at the root of the tongue, which is known to medical men as the larynx, it is called laryngitis; and this part is the common seat of this disease, from which it extends down the trachea, or windpipe, to the lungs. As long as the throat remains very sore, it is a pretty good evidence that the lungs are not affected.

THE SADDLE HORSE.

This disease may exist either in an acute, sub-acute, or chronic form. When acute, its management is simple and usually successful; but if it is neglected in this early stage, it not unfrequently proves troublesome, and in some cases leaves the animal permanently unsound, terminating in wheezing, whistling, roaring, or broken-windedness.

The symptoms of sore throat are easily detected by the ordinary observer. According to the intensity of the disease there is an accumulation of saliva in the mouth, clear, thick, and stringy, more particularly when the tongue is swollen; a stiffness of the head, the horse coughing upon the slightest pressure on the larynx; difficulty in swallowing, more particularly hard grain or hay, and a consequent refusal of food altogether; a short, hard cough; more or less copious discharges from the nose, as the disease advances; an accelerated pulse, frequently rising to ninety or one hundred pulsations in a minute; mouth hot, with considerable fever accompanying.

For treatment, apply strong mustard, mixed with water to the thickness of cream, to the throat, rub it well in, and repeat as often as may be necessary; or poultice the part with flaxseed meal for several days, and sprinkle on the tongue a teaspoonful of common table salt, three or four times a day, which in ordinary cases is all the treatment which will be necessary for the acute type of the disease.

The attention of the veterinary surgeon is more frequently called to chronic forms of this disease, in which, though no swelling of the parts is usually perceptible, a pressure upon the larynx at once excites a hard cough. In this stage of the disease much relief will be obtained by the application of a blister, prepared as follows: Pulverized cantharides (Spanish flies) half an ounce; of lard, one and a half ounces; mixed well, and as thin as may be desired with spirits of turpentine. This must be well rubbed in, and after it has acted thoroughly, dress with sweet oil or lard.

STRANGLES.

This is but another form or stage of laryngitis. The throat becomes enormously swollen, the swelling extending under the jaws and up to the very ears, threatening suffocation; then respiration becomes much disturbed; the flanks heave violently, and the breathing can be heard at a considerable distance; the animal begins to sweat from his frequently convulsive efforts to breathe, and, if not speedily relieved, dies a most violent death.

Life may be saved by the veterinary surgeon at this crisis by the operation of bronchotomy, that is, by opening the windpipe, and inserting a tube through which the animal may breathe instead of through the nose. This operation affords instant relief, and gives an opportunity to apply remedies to the diseased throat, which in a few days usually effect a cure, when the tube may be removed. The author has never lost a case where he has resorted to this operation.

The early treatment of this disease is to poultice the throat well with flaxseed meal, commonly called cake-meal or oil-cake, using salt upon the tongue as before. Mustard plasters are also very effective, and steaming the nostrils frequently affords relief. As soon as the swelling permits, it should be lanced; and when it has once discharged freely, the animal may be considered out of danger, provided proper care be taken to guard against a relapse. A seton applied between the jaws often relieves; but these cases are safer in the hands of a competent surgeon. Under no circumstances of this disease should the animal be bled.

Malignant or putrid sore throat, is fortunately but little known in the United States, the author not being aware of its existence in any portion. Cases presenting somewhat similar symptoms have been found upon examination to differ in a marked degree from those which accompany this form of disease as they are laid down in the works of foreign authors. A detailed description of this type of the disease is therefore deemed unnecessary in the present treatise.

CHRONIC COUGH.

This arises from various causes, and is present in a number of diseases. It is often symptomatic of some affection of the lungs and air passages; and it sometimes exists apparently as an independent affection, the animal thriving well, and retaining unimpaired his appetite and spirits.

If it arises from irritation of the larynx, or upper part of the throat, a few applications of mustard will be beneficial; if from worms in the stomach or intestines, treat as directed under the head of "Worms." If it exists without any apparent connection, or as the termination of disease previously existing, give every night in a bran mash one of these powders: of sulphate of copper (blue vitriol), digitalis (fox-glove), pulverized squills, nitre, and camphor, each one ounce; to be made into ten powders. Green food, as carrots, potatoes, turnips, or parsnips, should be given when procurable.

CATARRH.

This disease, commonly called a cold, is confined in ordinary cases to the lining membrane of the nose and neighboring parts; but in severe cases the inflammation sometimes extends down the air passages to the lungs, frequently resulting fatally. In the spring of the year this disease frequently appears in an epizootic form, when the symptoms are more alarming and the termination more generally fatal.

If the inflammation is confined to the nostrils, the membrane lining those cavities is reddened, a thin watery or mucous discharge from the nostrils takes place, accompanied with frequent sneezing; if the larynx is involved, there are cough, swellings underneath the jaws, etc.

Some authors recommend bleeding in this affection; but such an abuse of the lancet can do no good, and is often productive of much harm. If the symptoms are slight, one of the following powders given night and morning will be all that is required: of saltpetre two ounces; of pulverized Jamaica ginger one ounce; mixed, and divided into eight powders. If there is swelling under the jaws, poultice the throat with flaxseed meal; if much discharge from the nostrils, steam them well with boiling water poured upon bran. If the inflammation exhibits any tendency to extend down the windpipe, apply a blister all along the neck over the windpipe from the throat to the breast, giving one of the following balls night and morning; of nitrate of potassa and pulverized gentian root, each one ounce; Jamaica ginger and caraway seeds, each half an ounce; mix with molasses and divide into six balls. If the discharge from the nose continues, the animal losing flesh, and the appetite being impaired, give one of the following powders in the feed night and morning : sulphate of copper one ounce ; pulverized gentian root one and a half ounces ; pulverized ginger six drachms ; mix and divide into eight powders. Good wholesome food only should be given.

DISTEMPER.

All catarrhal affections are classed by horse-owners under the common head of distemper. Common catarrh, epizootic or epidemic catarrh, laryngitis, bronchitis, and all other diseases accompanied by nasal discharges, are regarded by horsemen generally as one and the same disease.

INFLUENZA.

For several years past a disease has been more or less prevalent in various sections of the United States, known to the veterinary profession as epizootic (epidemic) catarrh, or influenza. The symptoms of this disease are so various in different animals, no two being precisely alike, that a variety of opinions are current

QUIET ENJOYMENT.

concerning it and its nature, and, as a consequence, various other diseases are often confounded with it.

In the year 1855, this disease made its appearance in the stables of one of the largest omnibus proprietors in Philadelphia, and some nine horses died in about two weeks. These were supposed to have been foundered, and were treated for that disease. A careful examination, however, by a competent practitioner revealed the true nature of the disease, and under proper treatment the balance of the stock was saved. Shortly after the demand for veterinary surgeons was very great, and while they saved forty-eight out of every fifty cases, the farrier lost almost every case he attempted to treat, principally from his too common practice of bleeding and purging; thus reducing the system so low that nature became exhausted.

This disease is called by horsemen pink-eye distemper, and is by many regarded incurable, though the author knows of no disease that more readily yields to proper treatment, and in his own practice he has been eminently successful in accomplishing a cure. It commences with slight watery or thin mucous discharges from the nostrils; matter collecting in the inner corner of the eyes; eye-lid on the inner side of a very slight or yellowish red color; pulse feeble, with occasional paralysis of the hind extremities; sore throat; excessive debility; membrane of the nose much reddened; hard cough; heart sometimes violently agitated; flanks heaving; and feet sometimes hot; thus producing all the symptoms of founder.

For treatment, never bleed, as in nine cases out of ten, the animal dies. If inflammation runs high, as it sometimes does, use for several days the following: of tartar emetic and nitrate of potash, each two drachms, made into a ball with molasses and given at night. Give also in a pail of water one ounce of spirits of nitre twice a day; or, if more convenient, two drachms of the extract of belladonna (nightshade) dissolved in the water. When the inflammation is reduced. give one of the following balls night and morning : of pulverized gentian root and nitrate of potassa, each an ounce; pulverized Jamaica ginger, half an ounce; ground fenugreek seeds six drachms; mix with molasses, and divide into eight balls. In pure cases of debility (this being one of the serious symptoms of the disease), or in the early stages, previous to extensive inflammation being established, one of the following should be given twice a day :—sulphate of iron (green vitriol) two ounces; pulverized ginger one ounce; pulverized gentian root two ounces; mix with molasses, and divide into eight balls. In cases where the lungs are affected, give the following ball twice a day : of tartar emetic and pulverized digitalis (foxglove) each one scruple ; nitrate of potash three drachms; mix with molasses. Linseed tea, or oat-meal gruel should be given frequently. No hay should be given, unless the bowels are in good condition. If the liver is affected—which may be known by the yellow tinge of the mucous membrane, dung small and hard, horse lying on his side, and occasionally looking at his side as if in pain, with occasional fits of uneasiness—the following may be given, but must not be repeated ; of Barbadoes aloes three drachms, calomel and pulverized digitalis each half a drachm; make into a ball with molasses. In all these cases where there is soreness or swelling of the throat, the parts should be freely blistered ; and the sides also, if the lungs are involved. This mode of treatment has proved very successful in the author's practice.

BRONCHITIS.

The larynx (upper part of the windpipe), the trachea (windpipe), and the bronchial tubes (branches from the trachea into the lungs for the passage of air), are lined by one continuous membrane, called the mucous membrane, which secretes a thin mucous substance that always keeps the parts soft and moist. When this membrane becomes inflamed, the disease is named according to its location. If it is confined to the larynx (as has been before observed), it is termed laryngitis; if to the windpipe, trachitis; and if to the bronchial tubes, bronchitis. The trachea and bronchia are rarely diseased separately, the inflammation generally extending from one to the other. We shall therefore treat of bronchitis as embracing trachitis likewise. Even this disease rarely exists unmixed with others, in consequences of which it is often overlooked, or confounded with other diseases of a pulmonary character.

Bronchitis is generally preceded by a shivering fit; mouth hot, with more or less saliva; discharge from the nose; cough; sore throat; fever; short breathing; loss of appetite; accelerated pulse; and membrane of nose and eyelids reddened.

In treating this disease it is much safer to call in the veterinary surgeon, in consequence of the difficulty which the ordinary observer will experience in distinguishing it from other pulmonary diseases, and from the fact that the treatment varies with the changes that take place in the progress of the disease. It is not necessarily fatal; yet the most trifling neglect or mistake in treatment may make it so. The average loss, if proper treatment is pursued, is not more than five per cent. Resort should never be had to bleeding in any form which the disease may assume, although such treatment has been recommended by the highest authorities.

If much fever is present, give the following ball: of nitre two drachms; pulverized digitalis and tartar emetic each half a drachm; solution of gum arabic sufficient to make the ball. This may be repeated if the desired effect is not produced in twelve hours. Apply to the throat, sides, and along the spine, strong mustard mixed with water to the consistence of cream, which may be repeated as often as necessary. The fly blister is also recommended; but the author prefers mustard, as being so much quicker in its action. After the inflammation has subsided, give one of the following powders twice a day: of pulverized gentian root and nitre, each one ounce; pulverized Jamaica ginger, half an ounce; caraway seeds six drachms. This course of treatment is perfectly safe in the hands of any horseman, though it will not reach all stages of the disease; nor can any general directions be given better calculated to warrant a successful issue in these cases.

NASAL GLEET.

"Nasal gleet is the name here given to those discharges from the nose, which are commonly preceded by some inflammatory or catarrhal attack of the air passages, in particular those of the head; though there occur examples of their appearing without any such detectible precursors, originating, indeed, without any visible or apparent cause whatever; in most cases they are apt to continue long after all signs of inflammation have died away. Gleet is more likely to supervene after a chronic, than after an acute, attack of catarrh, and to show itself in an adult or aged horse rather than in the young subject. Sometimes the discharge comes from one nostril alone; more usually from both. Sometimes the submaxillary glands (glands under the jaws), remain tumefied, and sometimes they are not. The Schneiderian membrane (membrane of the nose) discolored by inflammatory action, has become pallid and leaden-hued, but is free from all pustular or ulcerative indications. The discharged matter varies in quantity and quality in different individuals, and even in the same horse at different stages of this disease. The ordinary gleet consists of a matter more mucous than purulent, remarkable for its whiteness, about the thickness of cream, and in some cases is smooth and uniform, in others clotty or lumpy; in other cases it is yellow, and appears to contain in its composition more pus than mucus. At one time it will collect about the nostrils, and become ejected in flakes or masses in pretty regular succession; at another time there is a good deal of irregularity in this respect, the running from the nose ceasing altogether for a while, as though the animal were cured, and then returning with double or treble force. Sometimes fetor is an offensive accompaniment of the discharge; at other times no fetor is perceptible. The health does not suffer in the least; on the contrary, it is one of the indications of this disease, that the horse eats and drinks, and has his spirits, as well as though he were quite free from complaint.

Formerly, these cases were considered to be evidences of glanders, and were called chronic glanders; many a horse having been destroyed under this mistaken impression. That a case of the kind might not turn to glanders, is, perhaps, more than can be asserted with certainty; but that, so long as

it continues gleet, it is not glanders, I am fully persuaded; and to show that it is not, I have been in more than one instance successful in bringing the case to a favorable issue."
[Percival's Hippopathology.]

The treatment recommended by veterinary writers has not been found successful in the author's practice; nor, indeed, do they themselves appear to have encountered any better fortune. That which has proved efficacious has, in all cases, been strictly tonic. Give the following powder night and morning for a month: of sulphate of copper (blue vitriol), half a drachm; pulverized gentian root, two drachms; pulverized ginger, one drachm; mix for one dose: or, give night and morning, mixed in the feed, half-drachm doses of powdered nux vomica (commonly called Quaker button). There is no danger in giving this preparation to a horse, provided he does not have water for some time afterward, say half an hour; and it very rarely fails.

PNEUMONIA.

By pneumonia, or inflammation of the lungs, is meant either a highly congested or an inflammatory condition of the lungs, arising from various causes, as high feeding, blanketing, close or badly ventilated stables, violent or extraordinary exercise, or sudden changes from heat to cold. Cold applied to the external surface of a heated animal drives the blood from the skin to the internal organs, often causing congestion of the lungs. Pulmonary diseases are more prevalent in the spring and fall, particularly if the weather be cold and damp.

This disease is generally ushered in by a shivering fit; the horse is sometimes attacked very suddenly; he refuses his food;

the respiration becomes disturbed, sometimes suddenly, at other times more slowly; legs, ears, and muzzle cold; cough sometimes present; staring coat; membrane of nose reddened or leadened-hued; the animal hangs his head in or under the manger, stands with his feet wide apart, remaining in one position with no inclination to move. The pulse varies very much; it is sometimes full and quick, at other times weak and scarcely perceptible.

In these cases auscultation is found of the greatest advantage in enabling one to detect to a certainty the true condition of the parts affected. If the attack is sudden, coming on after any violent exercise, and the pulse is quick, weak, and scarcely perceptible; by the application of the ear to the animal's side the case is decided, in the absence of all sounds, to be one of congestive pneumonia. In all these cases the less medicine which is used the better; they require the free use of the lancet, which must be promptly applied, or the animal dies. Blood must be taken until the animal begins to show symptoms of weakness; after which place him in a cool box with a pail of water, but nothing else, before him, the fresh air being all the medicine required. He will either speedily recover, or inflammation of the lungs will ensue. A second bleeding, notwithstanding the inflammatory action, is positively

THE RUNNING HORSE LEXINGTON.

injurious. As the disease assumes an inflammatory character, the breathing becomes more disturbed, the mouth hot, flanks heaving, and the nostrils expand and contract violently. Blisters must now be applied to the sides and breast, and those which will act quickly. The author prefers the following: of pulverized cantharides half an ounce; lard one ounce; croton oil twenty drops; linseed oil sufficient to make it liquid. Divide the following into five parts, and give one part internally every two hours: liquor ammonia acetatis twelve ounces; extract of belladonna one ounce; water one pint. If there is no improvement in twelve hours, give one scruple of white hellebore with three drachms of nitre every four hours until its action is manifest. This remedy, however, is a dangerous one in the hands of any but the qualified practitioner. Instead of it, the tincture of aconite may be used—indeed, it is one of the very best remedies. Take of tincture of aconite half an ounce to an ounce of water; give twenty drops on the tongue every three hours. Active purgatives should not be given; injections, however, are very useful. The horse should be kept on a low diet for a few days, as bran mashes, carrots, or green food; but no hay should be allowed, and a pail of water should be kept before him. This is regarded by the author in all inflammatory diseases as one of our best medicines.

PLEURISY.

By pleurisy is meant an inflammation of the pleura, or membrane covering the lungs and internal walls of the chest, without the lungs being involved in the inflammation; when, however, they partake of its inflammatory action, it is styled pleuropneumonia. The former disease rarely exists in a pure form; and as in a work like the present it is unnecessary to consider the delicately drawn distinctions between the two types, both will be treated as if they constituted in reality but one disease.

Pleurisy may exist in an acute or chronic form. The attack may be sudden, or gradual, the animal manifesting indisposition several days previous. A hard drive, over-exertion, exposure to cold, washing in cold water when warm, a fall, fracture of a rib, a punctured wound, &c., are all causes of pleurisy.

The horse manifests uneasiness; there is a violent heaving of the flanks, a looking round at his sides, with an anxious expression of the face; pulse quick and wiry; body, mouth, and breath hot; sweating in different parts of the body; a high state of nervous irritation, the animal pawing, lying down but rising immediately; a pressure against the side causes pain. A peculiar symptom is observable in this disease; the right fore-leg differs in temperature from the left, and such is the case with the hind ones; if the right fore-leg is warm, the left hind one will also be warm, and the others cold.

Experience proves that blood-letting in this disease is only opening the vein to let life escape; for if by this means we succeed in relieving the inflammatory action, the loss of blood so prostrates the system that the animal from pure debility becomes the victim of hydrothorax, or dropsy of the chest, living a miserable life for several weeks, perhaps months, to die at last from the accumulation of fluid in the chest. Bleeding, therefore, is uncalled for, and in fact is positively injurious. The early application of blisters to the sides is very important; and for this purpose the same preparation will be found serviceable as has been recommended in the case of inflammation of the lungs. The application of blankets saturated with hot water and kept round the body for several hours is very beneficial. Give one of the following powders on the tongue every hour :—of calomel one drachm; lactucarium (the juice of the common garden lettuce) two drachms; divide into three powders. In two hours after giving the last powder, give the following drench: liquor ammonia acetatis four ounces; sulphuric ether one ounce; tincture of aconite ten drops; water one pint. If no improvement takes place within six hours, give half a drachm of the extract of belladonna in a pail of water every three hours; continue this until the pupils of the eye dilate, or a favorable change otherwise takes place. If the pulse is weak, give two ounces of nitrous ether; one ounce tincture of opium; and half a pint of tepid water; but do not repeat the dose. The animal must be kept upon a low diet; no hay or corn should be given; carrots and green food may be used sparingly; give water frequently; injections of soap and water are necessary from the first attack. After the animal becomes convalescent, strong tonics must be given, as the case may even then terminate in dropsy of the chest. Nux vomica should be given in half-drachm doses in the feed at night; or half-drachm doses of the iodide of potassa dissolved in a pail of water may be given three times a day.

HYDROTHORAX.

Dropsy of the chest, or hydrothorax, is usually the termination of pleurisy in cases where bleeding or long-continued sedative medication has been practised. The fluid contained within the chest, if following an acute attack of pleurisy, is a beautifully clear, bright yellow fluid. In sub-acute cases there is considerable lymph floating in it, thus rendering it turbid. The quantity varies in different cases, from a quart or two to several gallons.

In this disease the animal stands with legs straddling; the breathing is short and quick, and as the water accumulates the respiration becomes more labored; pulse small and quick; staggering gait; breast, belly, and sheath swelled, leaving after pressure the impression of the fingers; if the ear is applied to the side, no sounds are heard.

No course of treatment can be suggested which would be likely to succeed in the hands of the amateur; this disease far too often proving fatal in the most skillful hands.

THICK WIND.

This disease differs in its action and effects from broken wind or heaves, though they are frequently confounded. It is characterized by a quickened respiration, in consequence of the obstruction existing in the air passages as the termination of inflammatory action. The capacity of the lungs is often very considerably diminished; the air-cells become filled up or obliterated; and the bronchial tubes become thickened; so that the same amount of atmospheric air cannot be admitted, thus giving rise to the quick, blowing action witnessed in this disease. "It is astonishing," says Mr. Spooner, "what great alteration of the structure of the lungs may exist, and the horse be still able to perform his accustomed work. I remember a horse that for some months worked in a fast coach, doing a stage of twelve miles daily in about an hour and a quarter. He was seized with inflammation of the lungs, and died in about sixteen hours.

On examining the body after death, it appeared that one half of the lungs for a long time past must have been perfectly useless, for the purposes of respiration, being so completely hepatized as to be heavier than water."

But little can be done in the way of treatment for a thick-winded horse. It is important to keep the bowels regular; and by feeding with good sweet provender some relief is usually afforded.

ROARING AND WHISTLING.

There are different stages of the same disease, arising from a thickening of the windpipe, or of the membranes of the larynx, rendering the passages smaller at the diseased parts. These diseases are generally the termination of neglected bronchitis, laryngitis, and all diseases of a pulmonary or catarrhal character; ulceration of the glottis (a portion of the larynx) is also a cause of roaring.

If these diseases are caused by tight reining, the bearing rein should be left off; if they arise from other causes, there is but little prospect of benefiting the animal, except in cases where the thickened parts are in an inflammatory condition, when relief will be afforded by the application of mustard plasters or fly blisters to the parts affected.

BROKEN WIND.

The cause of broken wind, or heaves, has never been satisfactorily ascertained; some writers attributing it to functional derangement of the digestive organs, others to rupture of the air-cells of the lungs, while yet a third class to a spasmodic action of the diaphragm, a muscle dividing the chest from the abdomen. In this disease there is a short dry cough, which is characteristic, and familiar to all practised ears.

THE ATTACK AND DEFENSE.

It is a singular fact, well known to all Western horse-owners, that this disease has no existence on the prairies of Indiana, Illinois, and other Western States; and broken-winded horses that have been taken to those sections soon get well, and remain so.

The symptoms of this disease are, a peculiar, double-bellows motion of the flanks; respiration quicker than natural; a short peculiar cough; and frequent passing of wind.

In its treatment the digestive organs should be kept in as healthy a condition as possible. The throat should be examined; and if by merely rubbing the sides of the throat a cough is excited, the chances for a cure are favorable; but if the windpipe requires a squeeze in order to produce the cough, there is little use in attempting a cure. Use upon the throat three times a week for five or six weeks the following salve well rubbed in: iodine ointment two ounces; blue (mercurial) ointment one ounce; mix well together, and make thin with oil. Give internally every night one of the following powders: of sulphate of copper and pulverized ginger, each one ounce; pulverized gentian root two ounces; divide into sixteen powders.

The benefits of this course of treatment have been very marked in the author's practice. In all cases no hay should be allowed, but wheat or oat straw will be found of great advantage.

DISEASES OF THE STOMACH AND INTESTINES.

INFLAMMATION OF THE STOMACH.

Inflammation of the stomach, or gastritis, is usually the result of swallowing poisons, or powerful stimulants. Mr. James Clark relates a case of death occurring from inflammation of the stomach in a horse in consequence of being drenched with a pint of vinegar; and another case where death was caused by giving a drench which contained half an ounce of spirits of hartshorn. A correspondent writing to the Turf Register in 1855, recommends the use of nux vomica, to destroy worms; to which the editor appends the following remarks:—"We must caution those not acquainted with the deleterious properties of nux vomica against giving that drug in large doses. Three nuts or buttons weigh eighty grains, and we have recorded evidence that sixty grains of the powder have killed a horse in a short time. Hoffman mentions that two doses, of fifteen grains each, proved fatal to the patient." The cause of these fatal terminations was doubtless some morbid condition of the stomach at the time the medicine was given. "I have known," says White, "a horse quickly destroyed by being drenched with a quart of beer in which one or two ounces of tobacco had been infused, and have seen other horses take much larger doses without any ill effects." The author has known cases where bots were supposed to have given rise to inflammation of the stomach.

The symptoms from poisoning are extreme distress and restlessness, with a perfect loathing of all food; the animal breaks out in cold sweats, lies down but rises quickly, and becomes quickly prostrated in strength; the pulse is quick and oppressed; purging may, or may not, exist.

The treatment will depend upon the cause of the attack, and should in all cases be intrusted to the hands of a competent practitioner, if one can be obtained. Where poison is suspected, it is better to give plenty of gruel, linseed tea, starch water, chalk water, with a couple of ounces of tincture of opium. The lancet should not be used, as the animal is already in a debilitated condition, which bleeding would only increase, thereby preventing the possibility of a speedy recovery.

INFLAMMATION OF THE BOWELS.

Enteritis, or inflammation of the bowels, called by farriers red colic, admits of three divisions: enteritis, or inflammation of the muscular coat of the intestines; peritonitis, or inflammation of the outer coat of the intestines and the membrane lining the cavity of the abdomen; and dysentery, or inflammation of the inner or mucous coat of the intestines.

The muscular and peritoneal coats are those usually involved in inflammation of the bowels; but the muscular is more frequently involved than the peritoneal coat. The causes of this disease are washing when warm, or swimming in a river, drinking cold water when in a heated condition, over exertion, costiveness, dry food such as hay with little water, worms, calcareous concretions, and metastasis.

The disease is sometimes preceded by a shivering fit; there is loss of appetite; hot skin; continued restlessness; mouth hot and dry; membranes of nose and eyes very much reddened; pawing; the animal lies down and gets up frequently, kicks at his belly, looks frequently at his sides; no cessation of pain; pulse hard, small, and wiry, often beating one hundred times or more a minute; respiration quickened; bowels constipated; dung small, hard, and dry; extremities cold; and the urine highly colored and passed with difficulty. As the disease progresses, the intensity of the symptoms very much increases. The animal is now covered with perspiration, which is succeeded by a chilly state; the pulse becomes quicker; the belly begins to swell; the entire system becomes prostrated, and the animal dies, frequently in the most violent manner.

These cases require prompt and active treatment, for the disease runs its course very rapidly, often terminating in the course of ten or twelve hours. If the costiveness yields early, the pulse becomes less frequent, soft, and full; the extremities regain a moderate temperature, attended with remission of pain, and the case will be likely to have a favorable termination. It is important that this disease should be distinguished from an attack of colic, since the symptoms of one very much resemble those of the other; the pulse, however, is the surest guide in distinguishing these diseases. The ordinary mode of treating colic would be highly injurious in the treatment of inflammation of the bowels.

In this disease copious bleedings are necessary. A large opening should be made in the jugular vein, and from six to eight quarts of blood taken, the quantity varying with the size and condition of the animal; the hardened dung should be removed by back-raking, after which tobacco-smoke injections are of great service; where these are not convenient, injections of soap and water may be used, or, what is better, an injection of two gallons of water with six ounces of tincture of arnica. One pint of linseed oil may now be given; and if the case be a very severe one, and likely to terminate in death unless relief be afforded, ten drops of croton oil may be added to the drench; but this last preparation should not be given except in very desperate cases, as of life or death. Aloes should not be given unless combined with opium; and even then this treatment is not advisable.

Blankets well saturated with hot water should be applied to the abdomen, and kept up for two or three hours; the legs should be well rubbed with cayenne pepper or strong mustard, and bandaged with strips of flannel; if there is no improvement in the course of four or five hours, give one drachm of chloroform in one pint of linseed oil, which may, if necessary, be followed in two hours by the following ball, mixed with molasses: one drachm of pulverized opium; half a drachm of calomel; and two drachms of linseed meal. The injections should be continued throughout; give linseed tea to drink, instead of water; soft mashes and new grass, if obtainable, may be given sparingly, but no hay, until the bowels are opened. The animal should not be worked for some days after recovery, as this disease is apt to return if he is put to work or exposed too soon. An attack of this character does not necessarily render the animal less useful or valuable after his restoration to health.

Peritonitis differs but little from enteritis. The horse is more affected with pain; the pawing, rolling, and kicking at the belly are most violent; the eye is wild in appearance; tenderness is evinced on pressing the abdomen; the pulse is full and throbbing; the dung is small and hard, and covered with a slimy substance. The same course of treatment should be pursued as is recommended for enteritis.

Dysentery (molten grease, or inflammation of the intestines), is often confounded with diarrhœa. It is sometimes accompanied with purging, but this is by no means an invariable symptom. The most common causes are irritation, translation or obstructed perspiration, and the administration of improper purging medicines, causing undue irritation, which terminates in inflammation. The animal usually evinces but little pain; the pulse is quick and small; there is sometimes purging, with great prostration of strength.

The belly should be well rubbed with the following wash: half a pound of strong mustard; four ounces of spirits of ammonia; and one pint of water. The following drink may be given every three hours until some improvement is observed, when it should be discontinued at once: of prepared chalk and tincture of ginger each one ounce; powdered opium one drachm; tincture of catechu half an ounce; tincture of red pepper two drachms; and one pint of water. Throw up injections of two ounces of laudanum in half a pint of water, frequently, and give thin gruel to drink. No blood should be taken under any circumstances.

DIARRHŒA.

This disease often arises in the absence of any inflammatory action upon the mucous surface of the intestines; and hence the distinction cannot be made by the ordinary observer between it and dysentery, if purging should be present. In order to obviate this difficulty we recommend only such remedies as are calculated to answer either case, without the possibility of doing injury by the administration of medicines

The causes of diarrhœa are over-exertion, exposure to cold, drinking freely of pump or spring water, and over doses of physic.

For treatment, give in one pint of thin gruel, one ounce of prepared chalk, half an ounce of tincture of catechu, two ounces of tincture of opium, and one ounce of tincture of ginger. Gruel, starch, or arrowroot should be freely given; good sweet hay is very advantageous, but no grass or bran mashes should be allowed.

OMAR PASHA, THE TURKISH CHIEFTAIN.

INORDINATE APPETITE.

Loss of appetite is soon observed and complained of by the horse-owner, and in too many instances gives occasion for improper medication. Some horses are particularly choice in the selection of their food, refusing that which is poor, or daintily and languidly picking it over. Horses sometimes eat slowly and daintily in consequence of weakness of the digestive organs; in such cases a handful of camomile flowers occasionally mixed in the food will be of great benefit. Boiled potatoes and the like will also be found beneficial in such cases.

The disease (for it is no less) of a voracious or depraved appetite arises from a morbid condition of the digestive organs, and is generally regarded by horsemen as a very desirable feature. The owner is greatly surprised, under such circumstances, that his animal does not thrive. A distinction must be made between a healthy and a morbid appetite. The former is indicated by the animal being ready for his food as soon as he comes in from work, and eating his allowance, if good sweet provender, with evident relish; but the latter is indicated by a constant craving for food and water, without regard to the quality of either, the animal oftentimes in addition to his usual allowance eating up the litter from under him, which is frequently in a very filthy condition. He is almost constantly craving water, and will drink even from a stagnant pool. We find him tucked up in the flanks, or carrying a big belly; his dung is often soft, slimy, and fetid; he stales largely, and his urine is often very foul; he is dull, lazy, and stupid, performing his work languidly or unwillingly. Such horses are more than any others subject to the disease next mentioned.

PALSY OF THE STOMACH.

In this disease, arising from a voracious appetite, the stomach becomes overloaded with food, and distended beyond its natural capacity. This is seldom observed until the symptoms are so plainly marked as not to be mistaken, developing in many instances the disease known as stomach staggers, which has been already mentioned. There are rarely any symptoms of acute pain; the pulse remaining in nearly its natural condition; respiration is but slightly disturbed; there is great heaviness of the head; the horse stands with the fore feet well under him, and appears to be weak in the knees; the membranes of the mouth and eyes present a yellow or orange appearance, indicating the liver as involved in the disease; the urine is highly colored; and in some cases there is paralysis of the eye, and often of the extremities.

The treatment required is much the same as in stomach staggers; in fact, this disease is the origin of the last named. Attention should be directed in the first place to opening the bowels, which requires a strong cathartic, made in the following manner: of Barbadoes aloes one ounce; of pulverized gentian root two drachms; pulverized ginger one drachm; mix with molasses. Give no food for at least forty-eight hours; a little water may be occasionally given. In twelve hours after the ball, give one scruple of calomel on the tongue, which may be repeated at intervals of twelve hours for two or three days.

RUPTURE OF THE STOMACH.

Rupture of the stomach or diaphragm is caused by the stomach and bowels being distended with food far beyond their natural capacity, or by an accumulation of gas in the stomach, as in flatulent colic. The diaphragm, or midriff, is often ruptured in cases of flatulence, as is the case also with the intestines. As nothing in the way of treatment can be offered in these cases, all speculation upon them is superfluous.

CALCULUS, OR STONY CONCRETIONS.

Calculous deposits are not unfrequently found in the stomach, intestines, bladder, kidneys, liver, brain, and in the glands, more particularly in the salivary glands; often giving rise to much difficulty, particularly when situated in the brain, salivary glands, or bladder.

Stones in the stomach and intestines of the horse are quite common. The author has seen several weighing from one to three or four pounds; and Mr. Spooner mentions one in his possession weighing little less than six pounds. There were found by the author in the stomach of a horse which died of colic, one hundred and fifty-one barrel nails, two buttons, and three small calculi. This horse belonged to a baker, and had been fed with the scrapings of the shop. The nails presented a very singular appearance, many of them being entirely covered with calculous deposits, and others covered with the same deposits on the heads and points, presenting a body with two heads.

The presence of these foreign bodies in the stomach and intestines occasions frequent attacks of colic, and sometimes produces inflammation of the bowels. Miller's horses are supposed to be most subject to these accumulations. These abdominal calculi generally have a metallic nucleus, are composed of the triple phosphates, and are generally round and smooth. When first taken from the intestines, they are of a brown or greenish color, but they soon become white. When a horse is subject to frequent attacks of colic, not occasioned by feeding upon corn, these accumulations may reasonably be suspected to be the cause.

HAIR BALL.

Hair balls are occasionally found in the stomach and intestines of a horse, generally accumulating around a metallic nucleus. There are several in the possession of the author where a piece of iron is the nucleus, and one where a piece of coal afforded the same basis. These balls occasion the same disorders, preceded by the same symptoms, and followed by the same results as the calculus. The animal may recover from a number of attacks of colic, and die at last from the same cause.

STRANGULATION OF THE INTESTINES.

On examining horses after death from an attack of colic, the small intestines are occasionally found tangled in a knot so as to cause a complete obstruction in the passages. This gives rise to colic pains, terminating in inflammation of the bowels and death. The small intestines being but loosely attached by the peritoneum, their outer covering, have free play in all directions, whence the tendency arises to these accidents; for the author believes them to spring from accidental rather than natural causes. There may be a simple twisting, or the intestine may be firmly tied into a knot.

There is another species, called intro-susception, or intra-susception, which is a slipping of one portion of the intestines into, or inside of, another portion, thus completely blocking up the passage. There are no symptoms by which either of these conditions may be known; and such cases are therefore treated as cases of ordinary colic, or of inflammation of the bowels, as the case may be. Where, however, such a condition of the parts exists, all treatment will be useless.

SPASMODIC COLIC.

This disease, called by farriers gripes, cramp, fret, &c., is a cramp or spasm of the muscular structure of the intestines, most generally of the small ones. The most common causes are the application of cold water to the surface of the body, drinking cold water when in a heated condition, costiveness, stones in the intestines, hair ball, strictures of the intestines, unwholesome food, &c.

The premonitory symptoms are sudden in their nature. The animal is first observed pawing violently, showing evident symptoms of great distress, shifting his position almost constantly, and manifesting a desire to lie down. In a few minutes these symptoms disappear, and the animal is again easy. But the same uneasiness again returns, increasing in severity until the animal cannot be kept upon his feet; the pulse is full, but scarcely altered from the normal standard. As the disease advances, the symptoms become more severe, the animal at times throwing himself with great force upon the ground as though he were shot, looking anxiously at his sides, sometimes snapping at them with his teeth, and striking his belly with his hind feet. The symptoms vary but little from those of inflammation of the bowels, the condition of the pulse and the remission of pain being the distinguishing features. The extremities are of a natural temperature; there are frequent but ineffectual efforts to stale, and a cold sweat bedews the body.

In this disease it is necessary to back-rake, and throw up the fundament injections of castile soap and water. Give internally two ounces of nitrous ether, one ounce of tincture of opium, and half a pint of water mixed, which may be repeated in twenty minutes with the addition of one ounce of tincture of aloes. Rub the belly well with mustard and water; if in half an hour there is no improvement, and no symptoms of inflammation are present, give of lactucarium half an ounce, of Jamaica ginger half an ounce, and one pint of the best rum or gin; shake well together, and give one-third with twice the quantity of water every hour until relief is obtained.

FLATULENT COLIC

This is an accumulation of gas in the stomach and intestines, occurring more often in the spring and fall than at any other season. Horses fed on corn are most subject to these attacks, in consequence of this kind of food fermenting readily in the stomach, more particularly when green. If the accumulation of gas thereby occasioned is not arrested, it soon swells the stomach and intestines to such an extent as to cause the diaphragm, or walls of the stomach to give way, and the death of the animal ensues. The author has known cases to terminate in death in less than half an hour from the observation of the first symptoms, so rapid is the course of this disease. The symptoms are the same in spasmodic colic, with the exception of the swelling of the abdomen. The same medicines are to be used, with the addition of from one to two drachms of chloride of lime in each dose,

SIR ARCHY, THE GODOLPHIN OF AMERICA.

according to the urgency of the symptoms. This, if given in time, will generally prove efficacious. Tincture of hartshorn and spirits of turpentine are recommended by some veterinary authors, and are excellent remedies; but as much injury has been caused by their use by inexperienced persons, the author would not advise their use since the animal may be killed by an improper administration of them.

WORMS.

Four kinds of worms are found in the horse, viz: the lumbrici, which very much resemble the common earth-worm in form; ascarides, so called from their supposed resemblance to a thread; tænia, or tape-worm, of which variety but little is known, as it is very rare; and, lastly, the persecuted bots, considered by farmers and horsemen the greatest of pests and the most dangerous of all the species.

The lumbrici are most generally found in the small intestines, where they sometimes do much mischief by their irritating effects. The author was recently shown a very remarkable specimen of these worms by his friend, W. W. Fraley, V. S. This specimen was some two yards long, consisting of a portion of the small intestines so completely filled with these worms as apparently to render it almost impossible for anything to pass through it, the worms having accumulated in thousands. These worms are from eight to ten inches in length, round and perfectly white. There appear to be two varieties of the lumbrici. The other variety is similar in form and length, but has numerous brown transverse lines, at about equal distances from each other, along its entire length.

The ascarides are found in the large intestines, and are white worms from one to three inches in length. It is a somewhat singular fact, that although these worms are usually found in the large intestines, their origin, apparently, is in the stomach of the horse. On opening horses after death, tumors are often found in the stomach, which upon being cut open will be found to contain either a thick whitish matter, or knots of small worms, from half an inch to an inch in length, of precisely the same appearance as that of the ascarides, and believed by the author to be identical with them.

The symptoms of worms are a rough, harsh, staring coat; irregular or depraved appetite; a whitish, or yellowish white, shining substance sometimes observable about the fundament, accompanied by a disposition on the part of the animal to rub the tail; breath occasionally hot and fetid; and in some cases a dry, short cough. The animal becomes poor in flesh and spirits.

Various modes of treatment have been adopted with but little benefit. The remedies which have become most popular are tartar emetic, calomel, turpentine, an infusion of Indian pink, arsenic, green vitriol, &c. That which has usually been found most successful in the author's practice is to give one of the following powders for three successive nights; of calomel three drachms; of tartar emetic one drachm; mix and divide into three powders. Twenty-four hours after, give the following purgative ball: of Barbadoes aloes six drachms; pulverized ginger two drachms; and pulverized gentian root one drachm. Oil of turpentine in doses of two ounces has been very highly recommended by some authors; but this the author regards as a dangerous remedy, from its tendency to produce inflammation of the stomach or bowels. Too many horses have been killed by its destructive agency to render its use advisable.

BOTS.

These are the larvæ of the gad-fly. During the summer months, when the horse is at grass, the parent fly is seen busily engaged in depositing its eggs upon the hairs of the animal in such places as are easily reached by his mouth. This seems to be an instinctive feature in this insect. The legs, shoulders, and body are the parts selected for this purpose. The gad-fly is seen hovering in an upright position when about to deposit her egg; she then darts upon the horse, fixing the egg to the hairs by means of a glutinous substance; she again prepares another, which is deposited in like manner, until many hundreds are observed covering the hairs of the animal. The rapidity with which these eggs are prepared and deposited is astonishing. They are taken into the mouth by the animal biting or licking himself or his mate, and are hatched upon the tongue, or taken into the stomach and there hatched. If the eggs are recently produced, they pass into the stomach before they are hatched; but if they remain for a considerable time upon the hairs, they are hatched by the warmth of the tongue, and they pass into the stomach, where they are developed. This fact may be easily and satisfactorily proven by taking the newly deposited egg in the hand, and then applying a warm fluid; when it will be observed that the egg is softened or dissolved, but does not produce the bot; whereas, if the egg be

COMMON GAD-FLY OR BOT.

old, it will hatch in the hand. The investigations of Mr. Bracy Clark, V. S., have thrown much additional light upon the natural history of these parasites.

The dread entertained of this species of worms by farmers and horsemen arises from the fact that so many useless books have been published, purporting to be guides to the farmer and horseman, many of which attribute the death of a majority of horses to ravages of the bot, and give as symptoms of their presence those which characterize inflammation of the bowels, kidneys, bladder, and the like. To this circumstance is to be attributed the vast distruction of life by drenching and physicking the animal for bots. Now, a rational view of the subject leads us but to one conclusion, viz., that the stomach of the horse is the natural habitation of the bot, and that it cannot be, or is not, developed anywhere else. This being the case, it is reasonable to suppose that inasmuch as the animal apparently suffers no inconvenience from their presence in his stomach, they were intended to serve some good purpose, rather than to do mischief. Indeed, without going to the extreme of asserting, as does Mr. Clarke, that bots are "always harmless," it may be safely asserted as the unanimous opinion of veterinary surgeons (farriers are not included), the world over, that they are comparatively harmless, and that when they do become injurious, it is almost always preceded by some morbid condition of the digestive organs. This may either arise from disease, or from enormous accumulations of bots, which are sometimes so great as to completely block up the pyloric orifice, or opening from the stomach into the intestines.

EGGS ON A HAIR.

EGGS MAGNIFIED.

There are no symptoms by which the existence of bots is indicated, except it be in the spring, when they pass from the horse by the fundament, assuming again the form of a chrysalis to re-produce the parent fly. As has already been stated, the symptoms of other diseases, as inflammation of the bowels, &c., are often assigned as indicating the presence of bots, but although bots may sometimes give rise to these conditions, it is worse than folly to jump at the probable cause in such cases and say that it is a case of bots because a horse looks at his sides and the like. When such an instance is encountered, no matter whether it arise from bots or not, the animal must be treated for the inflammation which is present. If we succeed in controlling it, and restoring the stomach to healthy action, the bots are no longer troublesome; but if, on the contrary, we commence drenching the animal for bots, the chances are that we shall kill him. Morbid conditions of the stomach will sometimes so incommode these little creatures as to cause them to escape from their unpleasant situation, which is commonly effected by perforating the walls of the stomach and allowing the fluids to escape into the abdomen, in which case no medical agent will save the animal's life. Fortunately, however, these cases but rarely occur. The author has met with but a solitary case in an experience of ten years where death could be attributed to the action of bots.

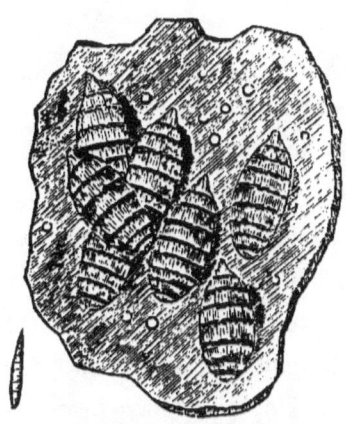

CATERPILLAR OR LARVÆ ADHERING TO THE LINING OF THE STOMACH.

We know, moreover, from frequent experiments that the

horse bot is more tenacious of life than even the cat, which is  popularly endowed with nine lives. The live bot has been immersed in spirits of turpentine, alcohol, nitric and muriatic acid, and many other equally powerful fluids, and yet he still adhered to life with marvelous tenacity. If, then, it were possible to detect the presence of bots by any marked symptoms, the attempt to remove them would certainly be hazardous to the life of the animal. The author has known cases of flatulent colic to be treated for bots, when, upon opening the stomach after the death which inevitably ensued, not a solitary bot was to be found. It will be borne in mind that in large cities, where horses are not indulged in a run at grass it is no unusual occurrence to find their stomachs entirely free from bots.

THE RED GAD-FLY.

CATERPILLAR OF THE RED GAD-FLY.

DISEASES OF THE LIVER.

Diseases of the liver are of very common occurrence in the horse, although the singularity of the internal structure of that animal renders it less liable to jaundice than the human being. The horse possesses no gall-bladder; instead of such a reservoir it has simply a gall-duct, called the hepatic duct, which enters that portion of the intestines called the duodenum about six inches from the stomach, so that the gall is emptied into the bowels as fast as it is secreted. Various opinions have been expressed touching this singular arrangement in the liver of the horse, any examination of which would be out of place in the present work. We proceed therefore to the mention of such diseases as come apparently under the above head.

INFLAMMATION OF THE LIVER.

Hepatitis, or inflammation of the liver, does not generally exist as a primary affection, though it is frequently found as a sympathetic one, being not uncommonly connected with epidemics, or epizootic diseases, particularly in that which is known to horsemen as pink-eye distemper.

The most common cause of this disease is a fullness of blood, or a plethoric condition of the system, in consequence of which too much blood is sent to the liver; want of exercise, and too high feeding, particularly with corn, are also causes of inflammation of this important organ.

The symptoms of this disease are more obscure than those of any other part, and the difficulty is materially enhanced by the inability of the animal to assist us with his tongue. Still, by close observation we can trace the symptoms with such a degree of accuracy as to render our treatment almost a certainty. The mouth and breath are hot; the extremities cold; the membrane lining the eyelids highly injected, presenting an orange-red appearance; the pulse rises from seventy to one hundred or more a minute, and is soft and full; the appetite lost; the animal looks wistfully and deploringly at his sides; lies down, but gets up again directly; the respiration at times is perfectly tranquil, at other times slightly disturbed, and at others again very much disturbed, and distressing to the animal—so that, in fact, the amateur cannot be governed by this symptom,—there is usually much tenderness of the right side; and the dung small, hard, and generally dark-colored.

In the acute stage the animal is generally in a state of plethora, in consequence of which a small quantity of blood may be taken to good advantage; but in the absence of plethora he must not be bled; a blister may be applied to the sides, or the application of creosote will be found serviceable. Injections of castile soap and water should be used occasionally until the bowels are opened. Give every four hours one of the following balls: of Barbadoes aloes six drachms; calomel three drachms; mix with molasses, and divide into twelve parts. Keep the body warm, and bandage the legs with flannel; turn into a loose box stall, where the atmosphere is pure. When convalescent, give one of the following balls night and morning: of sulphate of iron two ounces; pulverized gentian root one and a half ounces; pulverized Jamaica ginger one ounce; and pulverized anise seed one ounce: mix with molasses, and divide into sixteen parts.

JAUNDICE.

This disease depends upon an obstruction of the biliary excretions, causing a yellow discoloration of the mucous membrane, fat, ligaments, and other tissues of the body; it will oftener be found in connection with other diseases than distinct and independent of them, although it does occasionally exist in a pure or unmixed form, the symptoms of which are not at first observed by the horseman on account of their obscurity.

The lining membranes of the eyelids and lips are of a yellow or orange color, extending even to the white of the eye; the dung pale, small, and bally; bowels generally constipated; appetite lost or languid; the animal hangs his head, is dull and mopy, and becomes very poor in flesh.

In the treatment of this disease the principal reliance is upon calomel; two drachms of which made into a bolus with flaxseed meal and molasses should be given, followed in twenty-four hours by a purging ball. The animal should have moderate exercise daily; his body should be kept warm; and if there be pain in the right side, apply a blister; if necessary, the calomel may be repeated in scruple doses once a week.

HEPATIRRHŒA.

This is a rupture of the peritoneal coat of the liver, and hemorrhage from it. It occurs most generally in aged horses, and is always preceded by structural derangement, or disorganization which, from the obscurity of the symptoms escapes notice until it is too late for medical aid. The animal generally does his work as usual until within a few hours of his death, keeping in full condition, and presenting to the eye of his owner no appearance of disease. The symptoms are so gradual in their development as to escape observation until the peritoneum, or covering of the liver, gives way, or becomes ruptured, from the great distension of the liver, when the blood flows freely into the abdominal cavity, giving rise to the most alarming symptoms, and the horse often dies within an hour after he is first discovered to be ill.

The symptoms which are noticeable are suddenly developed, and generally appear immediately after eating or drinking. The animal will sometimes fall suddenly, and die in a few minutes, without having shown any previous indisposition; at other times the respiration becomes hurried, the belly begins to swell, the pulse becomes gradually diminished and very feeble, partial or general sweating takes place, the animal walks with a tottering gait, the membranes lining the eyelids, lips, and nose, become blanched, indicating internal hemorrhage, there is a vacant stare in the eye, with great prostration of strength, which soon terminates in death. Upon opening the abdomen, it is found filled with dark venous blood in a fluid state, and the liver is several times its natural size, and exceedingly tender. Where it is possible to detect the existence of the disease in its incipient stages, calomel would be the appropriate remedy, as it is as justly entitled to rank as a specific for the diseases of the liver of the horse, as it is for those of his master—man.

DECAYED STRUCTURE OF THE LIVER.

This also is a disease of common occurrence, though like the other diseases of this organ, the symptoms, from their obscurity, are not well understood by the veterinary practitioner, but little attention having as yet been paid to its investigation.

The first symptoms noticed are loss of appetite; surfeit; the being hide-bound; rough, staring coat; food passing undigested; stools of a clay color; prostration of strength; readiness to sweat; pulse quick but feeble; respiration hurried; sometimes violent purging, after which the animal usually dies.

Caution is necessary in the treatment of this variety of diseased liver. Bleeding must not be resorted to upon any consideration. In the absence of purging, give one of the following balls every other day: of calomel half an ounce; Barbadoes aloes one ounce; resin three ounces; mix with molasses,

INFLAMMATION OF THE BLADDER.

and divide into six balls. Upon the intermediate days give of sulphate of potash one and a half ounces; carbonate of potash one ounce; pulverized Jamaica ginger half an ounce; linseed meal two ounces: mix with molasses, and divide into six balls.

DISEASES OF THE URINARY ORGANS.

INFLAMMATION OF THE BLADDER.

Inflammation of the bladder, or cystitis, is a disease of comparatively rare occurrence in the horse, and generally is found in connection with other diseases. It is commonly supposed to occur more frequently in mares; although the author's experience has not confirmed this supposition.

The symptoms are continual emission of urine in small quantities; the moment it enters the bladder it is again expelled, but voided with much straining; pulse accelerated; pawing; the animal looks imploringly at his flanks; and upon passing the hand into the rectum, the bladder will be found contracted, and hard as a ball, being also hot and tender.

VIRGINIA MILL BOYS ON A RACE.

For treatment, back-rake the animal in the first place, and then throw up injections of water, adding to every gallon three ounces of tincture of opium. Give internally one and a half pints of linseed oil, to which may be advantageously added one drachm of chloroform. Bathe the loins with the following mixture : of strong mustard, a quarter of a pound; water, half a pint; hartshorn, two ounces: mix thoroughly together, and rub it well in. Give half a drachm of lactucarium three times a day; or, if more convenient, the extract of belladonna may be substituted. Give plenty of flaxseed tea; if the animal refuses to drink it, drench him with it. No hay must be given until twenty-four hours after he becomes convalescent. This is one of the most dangerous diseases to which the horse is subject.

RETENTION OF URINE.

This disease, technically known as spasm of the neck of the bladder, is found more frequently as an attendant upon other diseases than as an independent affection. It frequently occurs in colics as an accompanying symptom, thus misleading the ordinary observer in his judgment of the disorder.

The most common symptom is frequent but unsuccessful efforts to stale. This, however, must not be depended upon too strongly; as it will sometimes be observed in horses that are comparatively sound in these organs, particularly in those that have been well cared for. In such cases this temporary retention of urine arises from a dislike on the part of the animal of splattering his legs in voiding his water; hence he will often retain it in the bladder, though painful to him, until the litter is placed under him, when he at once stretches himself, and the urine flows freely and copiously. This fact has given rise to a superstitious notion among horsemen, that there is some peculiar virtue in the straw to cause this sudden cure; as a consequence, we frequently hear the remark, "Put some straw under him—that will cure him," etc.

If, however, retention of urine arises from disease, the straw possesses no magic charm to afford relief. In such instances the animal manifests but little pain, and rarely lies down. On passing the hand up the rectum or fundament, the bladder, which is easily felt, will be found very much distended with urine.

The services of a regular veterinary practitioner will be required in the treatment of this disease, as the bladder must be at once evacuated, which can in most cases be accomplished by means of an instrument called the catheter, which is not commonly found in the hands of any but the qualified surgeon. This desired evacuation can in some instances be produced by careful manipulation. Back-raking is very necessary in these cases, and injections of soap and water should be freely used. Unless the bladder is speedily emptied, it swells and bursts, causing a fatal termination. Fomentations of hot water to the abdomen, and pressure of the hand upon the bladder will be of assistance in enabling the animal to void the urine.

PROFUSE STALING.

This disorder, called also diabetes, is of frequent occurrence in the horse, and is attended with debility, impaired appetite, and sometimes loss of flesh. The causes are the improper use of nitre, saltpetre, and other powerful diuretics, as also unwholesome food, and the like.

The treatment is simple and effective; a great variety of medicinal substances being used in its abatement—as catechu, oak bark, gum kino, opium, chalk, etc. Either of these in moderate doses will usually check the copious flow of urine. Either of the following will be found sufficient: *uva ursi* (bear's whortleberry), powdered, two ounces; oak bark pulverized, four ounces; catechu pulverized, one ounce; opium pulverized, two drachms: mix either with molasses or honey, and divide into six balls, giving one every day. Or, the following may be used with equal advantage: opium pulverized, half an ounce; sulphate of iron, one ounce; gentian root pulverized, one ounce: mix with molasses, and divide into six balls—one to be given every day.

BLOODY URINE.

This disease, known also as hematura, frequently arises from strains across the loins, violent exercise, unwholesome food, calculous concretions in the kidneys, etc. It is not attended by symptoms of general derangement; the appetite is not usually impaired, nor is any marked degree of fever present. The color of the urine first calls attention, in voiding which the animal appears to strain slightly.

If the bowels are at all costive, injections should at once be thrown up the rectum; linseed tea should be given as a drink; mustard applications to the loins. Give internally one of the following once a day: of sugar of lead, one ounce; linseed meal, two ounces; mix with molasses or honey, and divide into eight pills; follow this for ten or twelve days, with one drachm of sulphuric acid in a pail of water to drink. Catechu, logwood, dragon's blood, oak bark, etc., have been used with advantage.

STONES IN THE KIDNEYS.

These concretions, which are quite common in the horse, are of a pale, dirty yellow color, elongated or conical in form, and much softer than any of the other varieties heretofore mentioned. "We have better evidence," says Mr. Blain, "than mere supposition; for urinary calculi (or stones in the kidneys), have been found in horses which have died with symptoms which might have been mistaken for very acute enteritis, or inflammation of the bowels. We may also suppose that the early accumulations would occasion irregular and diminished secretion of urine, followed at length by a bloody purulent mixture with the water, until more active symptoms should arise, and carry off the horse. Concretions within the kidneys might be removed in their early state by remedies tending to decompose them in the urinary pelvis. For this purpose we have mineral acids, of which the hydrochloric, as holding the silicious matter in solution, is to be preferred. The mineral acids pass through the body unchanged, being emitted with the urine in a state of purity."

A better opportunity is afforded us of discovering calculus in the urinary organs, than in any other parts; for an examination of the urine, when placed under the microscope, will enable us to detect its presence. When these deposits are ascertained, give in every pail of water which the animal drinks, two drachms of hydrochloric (muriatic) acid, which will in a short time be drunk with a relish by him.

STONES IN THE BLADDER.

These differ from stones in the kidneys in form and external appearance; presenting, in consequence of the constant washings of the calculus by the urine an uneven, or what is called a mulberry appearance; externally, it is of a reddish brown color. When these stones are quite large, very great inconvenience is occasioned to the animal.

Stones in the bladder may exist a long time before any perceptible symptoms of their existence are manifested. The urine is generally thick and of a whitish color, with frequent desire to void the urine, accompanied with difficulty and pain; the urine occasionally presents a bloody appearance; in some cases all the symptoms of colic are present, rendering it difficult to distinguish between the two disorders. If the pain is severe, the animal paws violently, kicks at his sheath, lies down, rolls, and gets up again quickly, sweats in various parts of the body, giving off the odor of urine.

For treatment, we should first attempt the dissolution of the stones, as recommended for stones in the kidneys, or we should remove them by the operation of lithotomy, which will be described under the head of surgical operations. If they are small, they may sometimes be extracted through the urethra, a process which is very easy in the case of mares.

DISEASES OF THE FEET AND LEGS.

CONTRACTION OF THE HOOF.

To horses that are kept in cities, or in stony sections of the country, this disease is one of the most common occurrence. In the middle and southern portions of New Jersey, and Ohio, and in many other sections where the shoeing of the horse is not called for except in frosty weather, contraction of the hoof is comparatively rare, in consequence of the feet being unfettered by that iron band, the shoe.

THE FAST-TROTTING STALLION, GEORGE M. PATCHEN.

This trouble is gradual in its approach; the first indication being a dry, brittle, unyielding hoof; the heels gradually becoming narrower, until they are painful. The hoof no longer accommodates itself to the soft structure within its limits, and, in consequence, the concussion is greater and the elasticity very much less. The parts therefore become bruised, and fever ensues, which still further facilitates the contraction of the hoof by absorbing its moisture; lameness follows as the natural and inevitable result. Upon an examination of the animal sweenie is decided upon by the horseman as the disease to which he is subject; a disease, by the way, which, we beg to say, the veterinary surgeon never yet has met.

The primary cause of this trouble is, undoubtedly bad shoeing, the preventives of which have already been fully unfolded.

Standing upon plank-floors has also a tendency to produce it, as it absorbs the moisture of the hoof, and renders it brittle and liable to crack. Traveling upon hard stony roads, with shoes that are beveled inwards, also predisposes the feet to this disorder.

The treatment must, necessarily, be slow in its operation; yet by careful management it is sure. The shoes must, in the first place, be removed, and the feet well poulticed for several days until the hoof and frogs become perfectly soft. The animal should then be carefully shod, as heretofore directed; apply daily, until the heels are fully spread, the following ointment; of rosin, four ounces; beeswax, four ounces; lard, two pounds; tallow, one pound; melt together, and, when cool, stir in four ounces of oil of turpentine.

CORNS.

The first effect of contraction of the hoof is to bruise the sensitive parts within their horny limits at that part of the foot formed by the crust and bar, causing lameness, which may be acute or chronic. These bruises are commonly called corns. The reason why this portion of the foot should be so severely bruised is obvious. The crust and bar forming a triangular space between which a considerable portion of the sensitive laminæ lie, this bar by its resistance of the encroachments of the crust, causes a twofold pressure upon the sensitive parts, acting much as a vice, and thereby diminishing the triangular space. Upon examination of the foot the horn is found hard, dry, and brittle, with a strong tendency to crack on very slight concussion. On removing a portion of the horn at the part of the foot indicated, the parts are found to be contused, sometimes slightly, and at others severely. In the latter case the feet are in such a condition as to require prompt attention, or a sloughing, or discharge of matter, may take place, forming a sinus, or pipe-like opening, through the quarter, sometimes passing through the coronet, and producing a condition, or disease, known as Quitter, which often terminates in permanent lameness and deformity.

When the lameness is of a chronic character, the poor beast, owing to his deprivation of speed, is compelled to suffer all kinds of barbarous treatment, such as roweling, setoning, etc., etc. As few believe corns to be of so serious a nature, the most are ready to attribute the lameness to a disease, or a supposed disease, which exists only in their disordered imaginations.

As symptomatic indications, it may be remarked that the horse extends one foot in advance of the other, and rests upon the toe, which causes a bending of the knee, with a hard, dry, brittle, and contracted hoof.

By way of treatment, the hoof, around the corn should be cut away so as to prevent pressure from the shoe; the corn should be well cut out, and burnt with a hot iron, butter of antimony, muriatic acid, caustic silver, or the permanganate of potash. He should then be carefully shod, and, if the frog is elastic, a bar shoe nicely fitted, with a perfectly level bearing, would be best; if, however, the frog is hard and unyielding, such a shoe may prove injurious. Flaxseed poultices frequently applied to the feet, together with the use of hoof ointment, will be found effectual; a run at grass without shoes will also prove beneficial.

QUITTER.

This is an ulceration, or formation of pus, between the sensitive and insensible laminæ, or inner parts of the wall of the hoof, generally situated on the inside quarter, forming sinuses, or pipe-like openings. Neglected corns often produce this disease, as also caulking or bruises from any cause.

The first appearance upon the foot on the approach of this disease is a hard conical tumor, hot, red, and smooth, which soon becomes soft, breaks, and discharges pus. A probe should first be introduced by way of treatment, pointing out the direction of the sinuses; an injection of sulphate of zinc, one drachm dissolved in a pint of water, should be thrown into the opening in the foot by the means of a small syringe, once daily, and the foot should be occasionally washed with castile soap and water. The early treatment should consist in poulticing with flaxseed meal for several days. If the case is very slow, use two drachms of the chloride of zinc to a pint of water; inject in the same manner; cut away all loose parts of the horn, which will facilitate the cure. Glycerine has also been used by the author with marked benefit.

THRUSH.

This is a discharge of a matter from the cleft or division of the frog, which occasionally produces lameness. It originates from a filthy condition of the stable, the animal being allowed to stand in his dung, or upon foul litter. Horses that are well cared for are rarely troubled with it. The symptoms are a rottenness of the frog, accompanied by a discharge of fetid matter. Lameness may, or may not, be present.

For treatment, wash the feet well with soap and water; fill the cleft with powdered sulphate of copper, and pack over it a little tow; remove the filth from the stall, and the animal soon recovers. An ointment may also be used, made of equal parts of pine-tar and lard, melted over a slow fire; when cool, add sulphuric acid until ebullition ceases, and it is then fit for use.

CANKER.

This arises from neglected thrush, often proving very difficult to manage. It extends from the horny frog to the sensitive frog, and sometimes to the navicular joint, involving the surrounding parts, and causing much alteration or destruction of the structures affected. It is by no means always a local disease, but is influenced by a morbid or unhealthy condition of the blood. The author's attention was once called to a case of four years' standing, in which all the feet were involved, and the value of the animal thereby so depreciated that he was sold to a shoeing-smith of Philadelphia for the sum of twenty-five dollars, his cost being some two hundred and fifty dollars. All treatment had failed up to that time; yet, notwithstanding the long resistance of the disease, it gradually yielded to constitutional treatment.

For treatment, all loose horn should be removed, that the parts may be properly dressed. If taken early, the following wash may be used with success; of nitrate of silver, half an ounce; water, one pint; shake well together, and use once a day. Or, the ointment of tar, lard, and sulphuric acid, recommended in cases of thrush, may be usefully applied. Should this fail, apply once a day the following: of castor oil, one part; collodion, two parts; mix well together. Give internally half a drachm of powdered *nux vomica* mixed in the feed, which should consist of green food, mashes, and a little hay. Corrosive sublimate in solution has been used with decided advantage; as also chloride of zinc, chloride of lime, butter of antimony, tincture of myrrh, sulphate of copper, glycerine, and many other preparations.

SCRATCHES.

This disease, called also cracked heels, generally arises from neglect, such as allowing the horse to stand in a filthy stall. It is generally confined to the hind feet, and consists in a swelling of the skin, causing in it one or more transverse cracks, which discharge a sanious (thin, serous, and reddish) matter at times; while in other cases the parts are almost dry but scurfy.

For treatment, wash well with soap and water; take a shaving, or other soft, brush, and make a lather of soap and water, with which mix a small quantity of powdered charcoal; rub this well in the fetlock, and let it dry, after which it can be rubbed off. Two or three applications are generally successful. The collodion and castor oil will also answer a good purpose; a physic ball should first be given.

GREASE HEELS.

This is the result of weakness in the capillary vessels of the feet and legs, and is often preceded by dropsical effusions, which frequently exist upon the leg as far as the hock or knee. Common-bred horses are supposed to be more liable to this disease, while thorough-bred are comparatively free from its attacks.

The principal causes are, doubtless, over-feeding and want of exercise; since we generally find the disease associated with a plethoric condition of the animal. As symptomatic, the skin at first is hot, red, swollen, and tender, and discharges a white offensive matter of a greasy feeling. As the disease advances, this discharge thickens into the form of tears, and becomes hard, presenting a grapy appearance. Abscesses are sometimes formed about the heels, causing the sloughing away of a large portion of them.

This disease requires constitutional, as well as local, treatment. Give internally for four days one of the following balls: of Barbadoes aloes, one ounce; pulverized gentian root, half an ounce; pulverized ginger, two drachms; mix with molasses, and divide into four balls. Follow this with half-drachm doses of *nux vomica* powdered; wash the parts well with soap and water, and apply flaxseed poultices, mixed with a solution of sulphate of zinc, until the inflammation is considerably reduced; then bathe carefully either with glycerine, or a solution of sulphate of zinc, or the castor oil and collodion wash. If the discharge is very offensive, use powdered charcoal and soap suds, allowing it to dry upon the legs; a solution of the chloride of lime may also be used; or a weak solution of corrosive sublimate is beneficial.

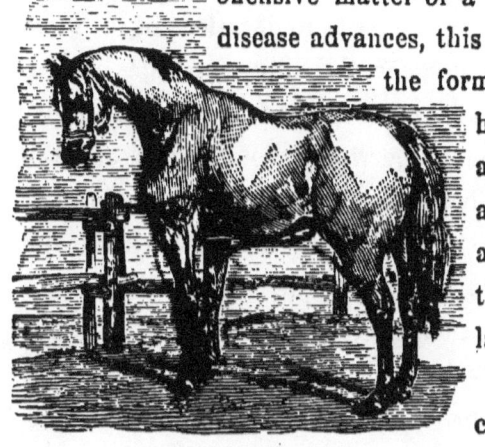

THE CHILDREN'S PET.

WATER FARCY.

This disease, together with *anasarca* and *œdema* may be classed under a common head, as they are but modifications of the same disease, which depends upon general debility for its existence. Two stages are, however, said to exist; one with inflammation, and the other without; one occurring in old horses, and the other in young ones. One important difference should be noted; the term *anasarca* is too extensive in its application to be properly associated with this disease, the term *œdema* being preferable to it, as having a local meaning and being more circumscribed in its limits.

As symptoms, the legs, belly, sheath, and other parts become swollen, and leave the impression of the fingers behind after pressure. In old horses such pressure rarely causes pain, inflammation being absent; but in young horses the legs particularly are hot and painful to the touch.

In this disease we must depend upon tonic and diuretic medicines; tonic, for the purpose of building up the system; and diuretic, to increase the secretions. The two should be combined as follows: of sulphate of iron, two ounces; pulverized gentian root, one ounce; pulverized ginger, half an ounce; nitrate of potash, one ounce; mix, and divide into eight powders, giving one night and morning, with good nourishing food, and allowing no corn. Or, the following will be found very useful: of sulphate of copper, one ounce; pulverized gentian root, one and a half ounces; pulverized ginger, half an ounce; nitre, one ounce; ground anise seed, six drachms; mix, and divide into eight powders, giving one night and morning. Hand-rubbing and daily exercise will be necessary.

WEED.

This is a disease similar to *œdema*, but makes its appearance above the hock, and extends downward. The skin is hot, and extremely sensitive to the touch; so much so that the animal throws the leg upward and outward as though to escape torture. The veins of the leg are full and corded.

For treatment, apply warm fomentations to the parts affected, and give a purging ball, followed by the powders recommended in the last disease.

CRACKED HOOF.

This disease, also called sand-crack, occurs only in the hoof that is dry, hard, brittle, and contracted. The hoof in a natural, elastic condition can be bruised, but not split up if double the force that splits the dry, contracted hoof is applied. This crack occurs most generally at the quarters, and almost always in the fore feet, they being almost alone subject to contraction. If the crack extends through the hoof it causes very painful lameness.

For treatment, the foot must first be carefully examined to see that no dirt has worked in under the hoof; the loose parts of the horn must be cut away; a pledget of tow, saturated with sulphate or chloride of zinc, or tincture of myrrh, should be applied, and a bandage carefully put on to keep it in place and keep out the dirt. As soon as the new horn has grown down a little, draw a line across the top of the crack with a drawing-knife or firing-iron, and apply a little tar or hoof ointment. If the crack is at the toe, a shoe with a band running across from the heels to a little below the coronet in front, and united by two screws, will often be all that is required, and the horse may be kept at work; but in quarter-crack it is unsafe to use the animal, particularly if it extends through to the soft parts. If the frog is in a healthy condition, which is rarely the case, a bar shoe, eased at the quarter, will be found beneficial.

SOLE BRUISE AND GRAVEL.

Accidents frequently occur to the feet of horses from their striking them forcibly upon stones and other hard substances. Pressure of the shoe upon the sole is the occasional cause of bruises of that part of the foot; and tender heels more frequently arise from bruises than from any other cause.

For treatment, if pus is secreted within the hoof—which may be discovered by the acute pain caused by a light tap of a hammer on that part of the hoof under which the matter is situated—the hoof must be cut through, that the matter may escape, as it will gradually work its way upward and make its appearance at the top of the hoof, thus rendering the treatment more difficult. After the matter escapes through the opening so made, throw in an injection of sulphate of zinc in solution, one drachm to a pint of water. For the treatment will be the same as recommended in quitter. Gravel sometimes works into these wounds, which must always be removed, and the parts carefully washed.

PRICKING.

This is an accident of too frequent occurrence, and happens in various ways, as by treading upon sharp bodies, such as broken glass, nails, etc., etc. It occurs more frequently, however, in shoeing, owing to the nail not being properly pointed, or, in some cases, from the iron not being good splits, one part turning inward and the other outward. These accidents are not always the fault of the smith, and he should not be unjustly censured for what he could not obviate. If such punctures are properly attended to, serious consequences rarely ensue. The practice of closing up the wound after removing the nail, glass, or other sharp substance cannot be too strongly condemned. It is doubtless in consequence of this senseless practice that so many horses are lost from lock-jaw, which does not generally make its appearance until the animal has apparently recovered from the wound; though upon an examination of the foot pus will often be found secreted within the hoof.

When a horse picks up a nail, or is pricked by the smith, a poultice should at once be applied to the foot, and kept on for several days; a cathartic ball should also be given, that the bowels may be in good order; after the removal of the poultice, apply the tar ointment, and no further trouble may be anticipated.

FALSE QUARTER.

This is an imperfect formation of horn at the quarter, which is generally of a lighter color than the other part of the hoof, and is divided by a seam from the top to the bottom. It is the result of injury from quitter and other diseases, rendering the heels weak, and requires the protection of a bar shoe, which should never bear upon it, as it may occasion lameness.

FOUNDER.

Founder, or laminitis, is an inflammatory condition of the *laminæ* of the feet, which are the most sensitive parts of these important appendages. Founder is said to be produced by various causes, such as hard driving, watering when warm, standing in a draught of air, or upon plank floors, and many others.

The author, however, views it in a different light, attributing its existence principally to one general cause, namely, contraction of the hoof, the causes before named being the immediate or exciting causes. This view is sustained by many facts. Founder does not occur in one case out of fifty in a healthy, open foot; nor are the hind feet often involved, as they are rarely in a contracted condition.

The symptoms are a full, quick pulse, from sixty upwards; accelerated respiration; the fore feet are hot and tender, the animal for relief throwing his body back upon the hind legs, extending the fore legs until he rests upon the heels, and sometimes lying down, particularly if the hind feet are involved; the animal also manifests much pain.

If the animal is in full condition, two quarts of blood should be taken from each of the fore feet; an active purging ball should be given, followed by one-drachm doses of belladonna made into pills every four hours; poultices of flaxseed meal should be applied to the feet for several days; injections of soap and water, also ought not to be neglected. By this treatment the animal is usually well again in a week, or even less; but if the disease is neglected until it becomes chronic, the animal will ever after remain unsound, though he may be rendered useful. From the alteration or disorganization of structure that takes place, there can little be done in the chronic stage except careful shoeing, which the smith should understand.

PUMICED FOOT.

This is called by horsemen a falling of the sole. It is preceded by founder, and is, in reality, one of the terminations of that disease, arising from the slow, continued inflammation of chronic founder, which causes absorption of the outer edge of the coffin bone, the latter thereby gradually losing its concave surface, and becoming convex. The sole, yielding to this gradual change, becomes flat, or, in some instances, convex. Very little can be done in such cases by way of treatment; yet by careful shoeing the animal may be rendered useful, although never sound.

CORINITIS.

This is an inflammation of the coronary ligament, situated within the upper part of the hoof and between the hoof and the hair. This ligament secretes the horn forming the wall or crust of the hoof, and when diseased ceases to perform its function, or performs it very imperfectly; as a consequence, the coronet, or upper margin of the hoof, is contracted, which causes the soft parts to bulge out in such a manner that it has often been mistaken for ring-bone. This contraction often causes lameness. The most frequent causes are, standing upon plank floors, hard driving, and the neglect to apply softening applications to the hoof.

For treatment, apply a flaxseed poultice for several days, and then a fly blister well rubbed in around the upper margin of the hoof; afterward use the hoof ointment once a day, until the coronet comes up full.

NAVICULARTHRITIS.

Coffin-joint lameness, as it is generally termed, is a disease of very common occurrence, and often troublesome to manage. This joint is formed by the union of three bones: the *os pedis*, or coffin-bone, situated immediately within the hoof; the *coronary*, or small pastern bone, the lower half of which is situated within the upper part of the hoof, called the coronet, and uniting with the *os pedis;* and the *navicular*, situated between and behind the two, uniting with both, and forming the navicular joint. This joint is protected against injury from concussion by the fatty frog, the sensible frog, and the horny frog, situated beneath it, and forming a soft elastic cushion on which it may rest. So long as the foot remains in a healthy condition, there is little danger of the occurrence of this disease. Even though the foot be strained very considerably, and a high degree of inflammatory action be produced, this disease will hardly arise, unless the inflammation becomes chronic. The author regards its origin as mainly due to a contracted condition of the feet, which, in fact, is the predisposing cause.

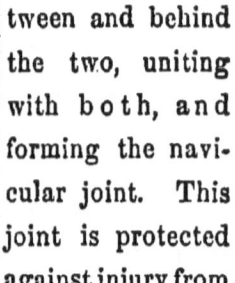

THE FAMOUS TROTTING MARE, FLORA TEMPLE.

Rarely, indeed, is navicular-joint lameness found existing in feet that have open heels and elastic frogs. If from any cause these frogs lose their moisture, they also lose their elasticity, and the foot therefore strikes the ground with a jar; inflammation of a chronic character sets in; the *synovia* (joint-oil) becomes absorbed; and *caries* of the bones is established, which destroys their articular surfaces and causes excessive lameness. Occasionally, owing to some new injury, acute inflammation sets in, causing new depositions of bone to be thrown out, and uniting the three bones together; which union is called *anchylosis*. This condition may be known by stiffness, and the animal walking upon the toe.

The symptoms of this disease have been confounded with those of another disease of the foot, which has been discovered by recent investigations. The horse is found to go lame upon coming out of the stable, which wears off after traveling some distance; one foot is observed in advance of the other when the animal is at rest; as the disease advances, the lameness becomes more frequent, until at last it is permanent. Various kinds of treatment have been resorted to, but with little success, such as blistering, firing, etc. Of late years, the frog seton has been introduced with very decided benefit. Should this, too, fail, there is no hope but in the operation of nerving, which should only be performed in certain cases mentioned under the head of neurotomy.

OSSIFICATION OF THE LATERAL CARTILAGES.

This is a transformation to bone of two projections of cartilage, or gristle, springing from each side of the coffin bone posteriorly, and known as the lateral cartilages. This disease was at one time called ring-bone, but the ring-bone of the present day is quite a different disease. It arises from concussion, and will rarely be found in any but contracted feet.

The treatment in these cases is only palliative, as the disease cannot be eradicated by any course of medical treatment. The first endeavor should be to expand the heels by applying poultices to the feet, together with the hoof ointment.

WIND GALLS.

Wind galls are puffy swellings about the joints, found above the fetlock on both the hind and fore legs. They are technically known as bursal enlargements, that is, a distended condition of the *bursæ* or synovial sacs, which contain the synovia, or joint oil. The animal suffers no inconvenience, apparently, from their presence upon his limbs, they evidently causing no pain.

It is seldom that any treatment is resorted to, except in the case of a very valuable animal. Blisters are commonly applied, but they are not attended with any permanent benefit. The application of cold water and compresses, secured by means of bandaging the legs, has proven the most efficacious.

SPRUNG OR BROKEN KNEES.

This trouble does not always result from an injury of the leg, or strain of the tendons; it is more often found in horses that have bad corns in the feet, or troubled with navicular disease, than in any others. The animal raising his heels to prevent pressure upon the tender parts, bends the knee, which bending becomes finally, from the altered position of the limb, a permanent deformity. Horses with sprung knees are unsafe for saddle purposes, owing to their consequent liability to stumble.

Respecting the treatment, it may be said that six out of every ten sprung-kneed horses will be found to have corns. If these be of recent growth, there is a fair prospect of straightening the limbs by removing the corns as directed under the head of that disease; by the removal of these the heels are brought to the ground, and the limb becomes straight. Under any other circumstances all treatment proves useless.

BREAKING DOWN.

This accident occurs in running, jumping, racing, etc. It is sometimes called a strain of the back sinews, and lets the animal down upon the fetlock, in consequence of a rupture of the ligament of the pastern. Horses meeting with this accident are of little value ever after, as they always remain weak in the fetlock. Unless the animal is quite young and valuable, the treatment would cost more than the animal's value. The French treat these cases very successfully by the application of instruments which keep the limb in its proper position until the parts have again healed and become strong. This is the only course to be pursued with any possible chance of a successful termination of the case.

STRAINS OF THE KNEES.

Strains of this joint occur in young horses while being broken into harness more often, probably, than at any other period of the animal's life. This results from the tenderness of the parts at that time, not one in twenty having them having arrived at maturity. These strains often prove troublesome to manage, and occasionally leave a stiff knee as the result.

Treatment—Bleeding from the plantar, or plate vein; warm fomentations to the part; when the inflammation is reduced, apply once a day for several days the following ointment: iodine ointment, one ounce; blue, or mercurial ointment, half an ounce; mix well together.

STRAIN OF THE HIP JOINT.

This occurs in falling, slipping, getting up, etc. The symptoms are a dragging motion of the limb; the lameness passing off after the animal gets warmed up, and returning upon his becoming again cool, the horse being then even more stiff and lame than before leaving the stable.

For treatment, apply cold water; a purging ball and rest are all that are requisite to effect a cure. Careful usage for some time after will be very necessary.

SHOULDER STRAIN.

This, which is of rare occurrence, arises from severe blows, or concussions; slipping so as to throw the legs apart forcibly; falling in the shafts of a heavily laden cart, etc. The symptoms are usually well marked; the horse is quite lame, both when walking and trotting; the leg drags with the toe on the ground, having an outward or circular motion.

Local bleeding is generally useful by way of treatment; three or four quarts may be taken from the plate vein, which runs down the inside of the leg. If, however, the animal is in a debilitated condition, bleeding should not be practised.

Foment the shoulder well with hot water frequently; a seton will often be found beneficial. After fomenting two or three days, use the following liniment: laudanum, one ounce; spirits of camphor, one ounce; tincture of myrrh, one ounce; castile soap, one ounce; alcohol, one pint. Or, sweet oil, one pint; spirits of hartshorn, three ounces; shake well together.

OPEN JOINTS.

These are generally the result of a punctured wound; the capsular ligament that surrounds the joint and confines the joint oil within its proper limits being thereby penetrated. These accidents are often attended with serious results, from the inflammation that is likely to arise from such an injury.

For treatment, efforts should first be made to close the wound, that the escape of the oil which lubricates the joint may be prevented. If the wound is small, it may be closed by means of a hot iron; if large, shave off all the hair around the opening, apply a piece of linen cloth well saturated with collodion, and bandage the part. Care must be taken to have the skin around the wound perfectly dry, or the collodion will not adhere. Shoemaker's wax, or common glue, applied in the same way, will frequently answer the purpose. The animal must be kept perfectly quiet, his bowels opened, and he be kept upon his feet for several days; if, however, the collodion adheres well, this is not of so much importance.

SWEENIE.

This imaginary disease has been the occasion of the infliction of much cruelty and unnecessary torture upon the horse. No respectable veterinary author recognizes any such disease. The symptoms which accompany its supposed existence are but sympathetic effects, or atrophy of the muscles of the shoulder. The attention of the horse-owner is directed to a wasting away or lessening of these muscles, which from want of action naturally become smaller or contracted; upon the animal's regaining the natural use of the limb, the muscles are again developed, as the muscles of the smith's arm by the constant use of the sledge hammer. Cases called sweenie are invariably the result of injury in some remote parts, as the knee, the foot, etc.— When the animal picks up the foot clear from the ground, it may be depended upon that the injury is not in the shoulder; if

THE EQUESTRIENNE.

however, the leg drags with the toe on the ground, the injury may be looked for in that locality. It is, however, more easy to decide a case of shoulder lameness than any other to which the limb is liable.

OSTITIS.

This is an inflammation of the bone, occasioning lameness of an obscure nature, and is one of the most difficult of all cases of lameness to detect. Where it occurs in the cannon bone, it is often mistaken for a thickening of the integuments.

Treatment—Cold bandages, lead water, rest, with daily half-drachm doses of iodide of potassa dissolved in a pail of water, will usually prove successful if the treatment be perseveringly adopted.

CAPULET AND CAPPED HOCK.

There are generally serous abscesses, produced by blows, bruises, strains, or injuries from any cause. Capulet is an enlargement at the point of the elbow, and is generally caused by lying on the heels of the shoe, which bruise the part. Capped hock is found at the point of the hock joint, and is usually caused by kicking against the sides of the stall.

By way of treatment, first open the part; if it contains fluid, which will be known by the soft elastic feeling, throw in with a syringe an injection of the tincture of iodine diluted with alcohol; a solution of the sulphate of zinc may in incipient cases answer the purpose. If fluid is not formed, blisters will often succeed. In cases of capulet, have the heels of the shoes shortened, or bind the feet at night to prevent injury.

CARIES OF THE BONES.

This is, perhaps, the most common of all the diseases to which the horse is subject, and its frequency can only be accounted for, by the abuses to which he is subjected. It generally arises from a low, inflammatory condition of the joints, these parts being principally affected; an ulceration of the heads of the bones is established, generally in young horses, which is called, from the destruction which it occasions, caries, or decay. It will usually be found preceding spavin, ring-bone, stiff back, and other anchylosed conditions of the bones, and can best be illustrated under the heads of Spavin and Ring-bone.

BONE SPAVIN.

This is a disease of such common occurrence that almost all horsemen think they fully understand its nature, pathological condition, and treatment. It is generally regarded by veterinary authors as a very serious injury, destructive to the utility of the animal, and very frequently reducing his value essentially in consequence of the blemishes. Where, however, there are no outward blemishes, as is the case in four out of every five spavined horses, the price of the animal is not affected, unless he is lame, since the disease is not discovered. There are, at this day, thousands of spavined horses traveling our roads, in not one of whom would the most experienced horsemen the world ever produced be able to determine the fact so long as the animal lives. In all such cases no external enlargement is found, but, on the contrary, the limb is clean and smooth. In the absence of enlargement, or spavin-bunch, as it is sometimes called, on the inside of the hock-joint, horsemen are unwilling to believe that spavin exists. The books, indeed, teach us to look there, and there only, for it; but the author's experience teaches him that the enlargement, where any exists, appears almost as often upon the front part of the hock as it does upon the inside.

Spavin generally arises from a strain, jar, or blow upon the hock-joint, causing an inflammatory condition of the cartilaginous cushions which cover the articular surfaces, or points of union, of each bone, or of the ligaments which surround the joints and bind the bones together; sometimes, indeed, both are involved. As this inflammatory condition is the exciting cause, spavin, or ulceration of the parts, speedily follows the neglect to remove it. When the inflammation is acute, the synovial fluid, or joint-oil, is soon absorbed; the cartilages of the joint are turned to bone, and uniting, one with the other, form one solid mass, destroying the elasticity as well as the mobility of the parts involved, and constituting what is called anchylosis of the hock-joint. This anchylosis, or union of bone, is not always general, there being in many cases but two, three, or four of the bones involved. When these changes are confined to the cartilage, external enlargement, or spavin-bunch, is never found. This the author calls spavin without any external indication.

When, however, the ligaments surrounding the joint are converted into bony substance, external enlargement in all cases exists. When a low, inflammatory action is found going on within the joint, it is an evidence of ulceration, in which, instead of new bone being thrown out, as in the acute stage, the natural bone is gradually decaying or rotting away. Hence arises the difficulty often experienced in the treatment of this disease.

As symptoms, the horse is very lame on leaving the stable, but when he is warmed up the lameness passes off; the leg is drawn up quickly with a kind of jerk; and there is a peculiar hard tread, which can only be distinguished by close observation. Where the bones are all united together, whether there is external enlargement or not, there is a peculiar twist of the heel outwards, which is more readily observed in the walk, and which the author has always found an infallible symptom of complete anchylosis.

Both spavin and ring-bone are incurable diseases. The lameness may be removed, but the disease, when once established, cannot, because the elasticity, mobility, and function of the joint are all destroyed in proportion to the extent of the disease. The spavined animal, therefore, comes down with a hard, jarring tread. The removal of the lameness depends upon perfect union or solidifying of the diseased bones. In the acute inflammatory cases, nature herself unaided works this change, and the animal recovers from the lameness with a stiff joint; but in the second, or ulcerative stage, assistance is required. We, therefore, endeavor to excite an active inflammation in the joint in order to overcome this ulcerative process, and induce new deposits of bone to be thrown out. Many modes have been adopted to secure the desired end, some of which are of a most barbarous character. Sharp instruments have been struck with considerable force into the joint, creating a tremendous fire, which soon checks the ulceration. This practice, although often successful, is unnecessarily severe, and cruel in the extreme. All kinds of caustic applications have been used, many of which have destroyed both the disease and the animal. Blistering the parts, the action being kept up for three or four weeks, often proves successful; firing is also practised; setons in the hock are frequently used with advantage. The following ointment is recommended; bin-iodide of mercury, one drachm; lard, two ounces; mix well together. Shave off the

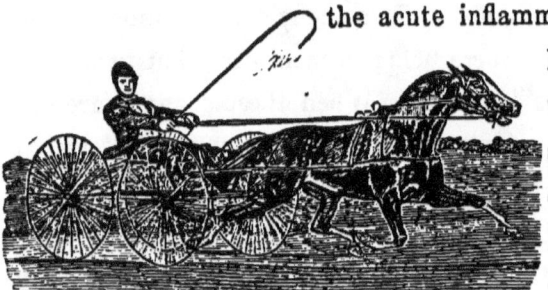

THE HIGH-BRED PACING MARE POCAHONTAS.

hair, and rub the part once a day for six or eight days; then wash the parts well with proof spirits. If the desired effect is not produced, repeat it.

RING-BONE.

This is a disease of the same nature as spavin, its locality alone giving it a different name; the same alterations of structure takes place; the same termination follows, and the same treatment is indicated. Contraction of the coronary ligaments is sometimes mistaken for ring-bone, and the poor beast is severely tortured in consequence. Contraction of this ligament produces a bulging of the soft parts around the coronet, causing the hair to turn downward and inward upon the hoof, giving it much the appearance of ring-bone. As in all such cases the heels are pressed close and painfully together, there is great necessity of distinguishing between the two before any application is made.

SPLINT.

This is an exostosis, or bony enlargement, arising from blows upon, or strains of, the splint bones, which are situated one on each side of the cannon bones and posterior to them. Splints are so common that few horses reach the age of eight years without having them, although they are not always visible to the eye at that period, having perhaps spread over a large surface of bone, or become flattened; which circumstance has given rise to the opinion among horsemen that old horses are not affected with splints. This, however, is a mistake; since a splint once formed is never afterward removed during the life of the animal. The nature of a splint is very similar to that of a spavin, but its course is somewhat different. When the injury is first received, the enlargement becomes quite prominent; but, as time advances, it generally disappears from view, even without the aid of man, spreading itself between the cannon and splint bones, thus lessening its size externally. Splints are not regarded as unsoundness, unless they cause lameness, which rarely occurs, particularly if they are situated near the middle of the bone; but if they are situated either at the upper or lower portions, or heads, lameness is almost always the result. This is easily explained; the bone, it will be observed, curves from above downward and outward, so that the lower extremity sets off from the body of the cannon bone; the upper heads, where it unites with the bones of the knee and hock, slant or bevel inward, and as the weight of the animal is thrown upon them, the upper heads are forced outward, while the lower ones are thrown inward. By this simple arrangement a rocking motion of these bones takes place, so that at the centre there is very little mobility, and if the injury is above, it causes lameness in consequence of tension; if below, from pressure; but, if it is in the centre, it seldom causes lameness at all, though the injury is greater.

When lameness occurs, the union of the bones should be hastened by increasing the inflammatory action; this is best done by active blistering, which soon removes the lameness.

CURB.

This is a swelling on the back part of the hock joint below the cap, generally arising from a strain, or breaking down of the hock. Some horses naturally have what are called curb-hocks, though they are not always attended with any serious disadvantages. There is a predisposition to weakness, which renders them suspicious.

If the curb arises from recent injury, a little blood may with advantage be taken from the sephena vein running up the inside of the thigh; cold water applications should be kept upon the parts; cloths wet with tincture of arnica, half a pint to a gallon of water, are very useful; or, the following ointment will be found of service: dry iodine, one drachm; iodide of potassa, one drachm; lard, one ounce; mix well together, and use once a day.

STRING HALT.

This disease has never been very satisfactorily accounted for by veterinary authors. It consists in a sudden, spasmodic raising of the hind limbs, though it is said to have occurred in the fore legs. The author has found, upon an examination after death of the hock-joint of several animals affected with this disease, that a little roughness from exostosis existed on the *os calcis*, or bone forming the cap of the hock, where the *perforans tendon* plays over; in other cases the tendon has been found almost entirely surrounded with a bony case, which interfered very materially with its action. He is inclined, therefore, to regard these as the general causes of the disease.

No treatment as yet practised has proved successful; though there are recorded isolated cases of spontaneous cure.

BLOOD SPAVIN, BOG SPAVIN, AND THOROUGHPIN.

These constitute one disease, occasioned by an over secretion of joint oil in the hock joint, which causes a distention of the capsular ligament, or *bursa*, presenting soft puffy swellings about the joint. Blood and bog spavin appear on the front and inside of the joint; while thoroughpin extends through from one side of the joint to the other. These diseases are so common and so well marked as not to be easily mistaken. The causes are violent exercise, throwing the animal upon his haunches, running, jumping, etc.

As it seldom causes lameness, treatment is rarely needed; if requisite, blistering, bandaging with compresses, and rest are the most successful.

FRACTURES.

Experience has established the fallacy of destroying every horse that meets with a fractured limb. Fractures may occur in any bone of the body, and yet a perfect union of the parts may take place, provided the fracture is a simple one; compound fractures, even, are occasionally united.

For treatment, the animal should first be placed in the most comfortable position, and the parts adjusted as nearly as possible, retaining them by proper bandages, splints, etc. Should the fracture be in the small or lower part of the leg, sole leather, softened in water and moulded to the limb, retaining it in place by bandages, forms a very good splint.

Fractures of the skull sometimes require the operation of trephining, (explained under the head of SURGICAL CASES,) in order to replace the parts perfectly; after which the bowels should be opened, and the animal kept on moderate diet.

Fractures of the *pelvis*, or haunch bones, will, in nine cases out of ten, become united by proper management, no matter how bad the crushing, and the animal may again be rendered serviceable. The author never hesitates to treat fractures of these bones in horses that are of sufficient value to warrant it. Indeed, union of the parts in such fractures will often take place, even if the animal be turned into a field without any treatment; though, perhaps, more deformity will be left than if proper care had been exercised. The horse, if active and high-strung, should be kept upon his feet by tying up the head short for several days, and then the slings may be placed under him; if this is done at first, the animal being full of fire throws himself off his feet, and all efforts to remedy the fracture will prove a failure. From six to eight weeks, according to the age of the animal, are necessary to complete the union of the parts.

Some practical knowledge is requisite, in order to discriminate cases of fracture of the limbs that are likely to be successfully treated; but fractures of the haunch bones rarely fail to unite, with proper management. The animal should be kept on bran mashes, gruel, and green food during the treatment.

DISEASES OF THE HEART.

Diseases of the heart are less understood by the members of the veterinary profession generally than any other class of diseases (with, perhaps, one or two exceptions,) to which horses are subject. This want of information in this country, is attributable to the comparative infancy of veterinary science, the obscurity of the symptoms by which these diseases are characterized, the consequent confounding of them with other diseases, and to the comparative silence of veterinary authors upon this important subject.

Diseases of the heart in this animal are not suspected by the farrier, (shoeing-smith) or horseman; yet they are by no means of unfrequent occurrence. During the session of the Veterinary College of Philadelphia for 1859-60, the author had then opportunities of presenting to the class well-marked cases of disease of this organ, as also one very interesting case of rupture of the heart, or rather of the aorta, or great artery leading from the heart, at the point where it leaves that important organ. The latter case was that of a bay mare which had been used in an oyster cart; she ate her feed at night as usual, in apparent good health, and was found dead in her stall the next morning.

PERICARDITIS.

This disease, as its name implies, is an inflammation of the pericardium, the bag or sac which surrounds the heart, and known to butchers as the heart-bag. After death arising from pleuritic affections effusions are quite commonly found within this sac, which are attributed to the sympathy existing between the pericardium and the pleura. The fluid is sometimes of a bright yellow color, while at others it is of a turbid character with considerable lymph floating in it, which collects in a mass forming a thick layer upon the internal surface of the sac, causing considerable thickening of its walls, and extending over the heart in like manner; adhesions between the two sometimes take place. Percival mentions an instance in which this collection was converted into a substance of the nature of gristle of considerable thickness. This disease rarely exists alone, but is of a secondary character.

The attendant symptoms are palpitation of the heart, quickened respiration, sometimes accompanied with a dry cough, with a pulse quick, rising to sixty or seventy a minute, full, hard, and strong. "Mr. Pritchard, V. S., Wolverton," says Mr. Percival, "with laudable zeal for the promotion of our art, so long ago as the year 1833, furnished the veterinarian with some practical communications on this subject, which we shall find it advantageous to revive upon the present occasion. His observations relate particularly to the type termed *Hydrops Pericardii*, which implies the stage of pericarditis when effusion is likely, or has taken place, and the membranous sac is supposed to contain watery fluid, and probably lymph as well. The symptoms of this affection, apart from pleurisy and pneumonia, Mr. Pritchard informs us, are well-marked. They are palpitation of the heart, the carotid arteries (passing up the neck) beating forcibly and being readily recognized in applying the finger to their course in the neck. There is a good flow of blood through the jugulars; a copious return of blood through the neck, when the state of the pulse is considered; the surface of the body and the extremities are warm; and these latter symptoms continue within one or two hours of the horse's death. * * * In addition to the above symptoms, there is such an expression of alarm and anxiety in the countenance of the animal as no other malady produces."

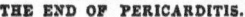

THE END OF PERICARDITIS.

There is no treatment as yet known by which this disease can be reached.

CARDITIS.

This is an inflammation of the muscular structure of the heart comparatively rare, or at least supposed to be so.

In this affection the animal will be found lame, generally in the off fore-leg, but upon examination no cause will be found sufficient to account for it. This lameness may appear and disappear several times previous to the attack's manifesting itself in a more positive form, leaving the impression that the lameness was rheumatic. We next find the animal refusing his feed; his heart palpitates violently; he occasionally gasps, and gnashes his teeth; pulse full, hard, and quick; there is a wild expression of the eyes; respiration quickened; mouth hot and dry; and the temperature of the legs varies from moderate to cold.

For treatment cold water should be frequently given; take one drachm of white hellebore, and divide it into five powders; give one of these on the tongue every three or four hours. Bleeding has been recommended; but the author has not witnessed any advantages from it, and therefore would on no account advise it.

ENDOCARDITIS.

This disease, called also palpitation of the heart, or, more commonly, thumps, is an inflammation of the lining membrane of the heart, and is generally associated with pericarditis; the inflammation readily extending itself from one part to the other in consequence of their proximity.

The symptoms are a violent palpitation of the heart, which

can often be observed at the distance of several yards from the animal; pulse full and hard, but not quickened. Although this disease is regarded as incurable, we can still palliate the symptoms so as to allow of the animal's return to work the next day.

For treatment, give one of the following powders every three hours; of nitrate of potassa one ounce; pulverized digitalis two drachms; mix, and divide into five powders. Subsequent attacks may be warded off by keeping the bowels regular.

DISEASES OF THE HEAD.

OSTEO-SARCOMA.

This disease, called commonly Big Head, is not mentioned by veterinary authors in Europe, and so far as the author can learn, seems to be peculiar to the Western and Southern States. It appears, from the rather unsatisfactory accounts at the author's command, to originate in the osseous, or bony, structure of the face. The bones become much swollen, and are represented as presenting a soft, spongy, or cellular appearance, the cells being filled with a substance like jelly. This appearance, however, does not correspond externally with several specimens in the author's possession, in which the external surface of the bones appears to be perfect, but very thin, and very much enlarged.

The symptoms are a swelling of the bones of the face from the eye to the nose; puffy swelling about the limbs; stiffness about the joints; pulse slightly accelerated, and soft; coat rough and staring, with considerable debility.

The treatment usually practised has been to make an incision through the skin and insert a small quantity of arsenic into the wound; or else to score the face with a red-hot iron; which latter mode is said to have effected a perfect cure in many cases. Neither of these operations, however, strikes us as being very scientific. The author's friend, G. W. Bowler, of Cincinnati, Ohio, has had some experience in the treatment of this disease, and has been very successful. The course pursued by him is to rub the swollen parts well once a day with the following ointment: of mercurial ointment one ounce, and of iodine ointment two ounces; mix well together for use. Give internally at the same time one of the following powders night and morning: calomel one ounce; iodide of potassa two ounces; pulverized gentian root one and a half ounces; to be made into twenty powders. The animal must be kept in a dry, well ventilated stable, and the body kept warm so long as this medicine is given.

INFLAMMATION OF THE BRAIN.

This disease, known also as phrenitis, or, more generally, mad staggers, arises from various causes, such as blows, overfeeding and little exercise, too tight a collar, etc., etc.,

A heaviness of the head is first noticed; an unwillingness to move about; the lining membrane of the eyelids much reddened; appetite indifferent or lost; a peculiar dullness of the eyes; and finally, delirium or madness. The animal becomes unmanageable; beslavers all that comes within his reach, whether man, horse, or anything else; and plunges violently about the stall, or wherever he may chance to be.

As this disease is occasioned by a determination of blood to the head, it is necessary to use the lancet; this should be done freely, and that too before the delirious stage comes on, otherwise it cannot be done properly or beneficially. Cloths wet in cold water should be applied to the head; or, what is better, bags of broken ice. Open the bowels with the following, made into a ball: Barbadoes aloes one ounce; pulverized ginger one drachm; pulverized gentian root two drachms; mix with molasses sufficient to form the ball. Give also injections of castile soap and water. Give no food for twenty-four hours; but small quantities of water may be frequently given. After the recovery of the animal he should be fed very sparingly, and not exposed to the hot noonday sun.

If the occasion of the attack be a tight collar, the remedy is simple and easy; if from over-feeding, the quantity of food should be lessened; but little is to be expected by way of treatment.

MEGRIMS.

This is a sudden determination of blood to the head, generally attacking horses while at work, or in harness upon the road. Those of a plethoric character are most subject to these attacks.

The horse suddenly stops in the road, shakes his head, and sometimes goes on again; at other times he falls in a state of unconsciousness, the whole system appears convulsed, with the eyes wild in appearance and constantly rolling.

Bleeding upon the appearance of the first symptoms gives almost immediate relief; after which the bowels must be opened, for which purpose give one and a half pints of linseed oil, or the aloes ball will answer; bran mashes should be given for a few days. These attacks may be prevented in the case of horses subject to them by moderate feeding and driving, and in warm weather by keeping the forehead shaded by a canvas or cloth hood elevated on a wire framework about two inches from the forehead so as to protect the brain, and admit a free passage of air between the two. The author believes that he was the first to introduce this hood, which can be attached to the bridle, and made as ornamental as may be desired. The use of hoods of this kind in very hot weather would prevent the frequent falling of horses in our streets from over-heating; as the heat of the sun principally affects the brain in all these cases.

VERTIGO.

This disease generally arises from water in the cranial case, causing pressure upon the brain. The animal is generally attacked in harness, as in the preceding disease; this arises from the fact that the exercise causes the vessels of the brain to become more active, fuller, and more distended with blood, and consequently there is greater pressure upon this sensitive organ.

The symptoms are similar to those of megrims, with, perhaps, the addition of rearing, dropping suddenly as though struck with death, and rising in a few moments as if nothing had happened, etc.

The treatment mainly consists in keeping the bowels in good order; working moderately; giving no corn, and but little hay.

EPILEPSY.

This disease takes its name from the suddenness of its attack. The animal is apparently in a perfect state of health, when suddenly he falls to the ground, generally (as in the two preceding cases), while in harness, without any manifest cause. He remains in this condition for a short time, and then appears as well as ever; although occasionally a considerable degree of stupor is manifested for some time after.

It may be occasioned by blows, wounds, and other injuries about the head; water in the brain; tumors; violent derangement of the nervous system; worms; constipation of the bowels; plethora, etc.

The same course of treatment should be pursued as in vertigo; these diseases in their symptoms, causes, etc., being so intimately connected as scarcely to be distinguishable from each other.

STOMACH STAGGERS.

This disease arises principally from over-feeding. The animal appears dull and sleepy, with a disposition to pitch forward; stands with his head resting against a wall, manger, or the like, or, if at pasture, against a tree; if he is led out of the stable, this will be observed as an involuntary action, in consequence of which the head is often much cut and bruised by coming in contact with hard or rough substances; there is constipation of the bowels; pulse scarcely changed from the usual standard; as the attack is severe, the breathing becomes more and more labored.

Blaine regards these symptoms as the first stage of mad staggers; but this the author deems a mistake, as animals that die from this disease, having presented the above symptoms, scarcely have any very marked change in the cerebral region, or the brain.

From the mode of treatment recommended by European authors of high repute, the author infers that the attacks are less severe in this country than in Europe, or else that the severe treatment there practised is more injurious than the disease itself. The whole cause of the disease being apparently in the distended condition of the stomach from the presence of undigested food, all food should be removed from the manger, and none given for forty-eight hours. Give internally the following ball: Barbadoes aloes one ounce; pulverized ginger two drachms; croton oil six drops; mix with molasses, and give in the usual manner. Injections of soap and water should be given, until the bowels are opened; or, what is far preferable when convenient, tobacco-smoke injections. Two drachms of the extract of belladonna dissolved in a pail of water, given to drink once a day for a week, will prove beneficial. Bleeding in these cases is, as a general rule, unnecessary and uncalled for. Food should now be given very sparingly; and no corn should be given at any time to the animal after such an attack, in consequence of its tendency to heat the blood, and produce a plethoric condition of the system.

HAYING SCENE.

DISEASES OF THE EYE.

AMAUROSIS.

In this disease, called also Gutta Serena, or, more generally, Glass Eye, we find the eyes bright and clear, with a peculiar glassy appearance about them not observed in an eye where vision is perfect; although no alteration in the structure of the eye has taken place, yet the horse is partially or totally blind. A mere examination of such eyes would not enable us to pronounce upon the blindness of the animal; but if he be taken from a dark stable to a strong light, it will readily be detected, as the light causes no change to take place in the pupil.

This disease is regarded as paralysis of the optic nerve; in some cases yielding readily to medical treatment, and in others proving incurable. Horses are often sold with this disease upon them as perfectly sound, and the first intimation which the purchaser receives of his horse's being blind is his running against a wall-fence, post, or any thing that may chance to be in his way. It sometimes makes its appearance very suddenly; occasionally it exists in a temporary form as a sympathetic affection, as in apoplexy; it also at times occurs during the period of gestation, etc.

Constitutional treatment only is likely to succeed in these cases. A physic ball should be given to open the bowels, composed of Barbadoes aloes six drachms; pulverized ginger one drachm; pulverized gentian root two drachms; mix with molasses. After the ball has operated (which should be in twenty-four hours, if the aloes are good), give morning and evening half a drachm of nux vomica mixed in the feed. The author has never witnessed any beneficial results from bleeding, although it is recommended by some writers.

INFLAMMATION OF THE MEMBRANA NICTITANS.

This affection is commonly called the haw, or hooks. The membrane affected is somewhat triangular in form, concave on the inner side, and convex externally. It is mainly composed of cartilage, or gristle, and is situated between the eye ball and the side of the orbit, at the inner corner of the eye. In a perfectly healthy condition but a very small portion of this membrane is visible; but when in a state of inflammation it bulges out very considerably. A portion of the membrane covering it becoming, as it were, folded upon itself presents a hook-like appearance, which has been regarded by farriers as a foreign substance, to which the name of "hooks" has been given, and its removal with the knife recommended by them. It so happens, however, that this membrane is placed in the eye, or attached thereto, to serve a useful purpose: that of cleansing the eye from dirt, or any foreign substance that may chance to get into it, which is accomplished by throwing it over the ball of the eye, and removing any obstruction. Injury must result from cutting away any portion of this membrane, as its function is in part destroyed; since the animal can no longer throw it over the ball of the eye with the same facility as before the operation was performed. In point of fact, wherever the hooks, as they are called, are cut out, it will be observed that whenever any foreign substance gets into the eye, the animal makes a spasmodic effort to throw this membrane over the eye ball, often failing to accomplish it; and thus the eye is rendered more liable to injury ever after.

Whenever this membrane becomes tumefied, instead of cutting it out, open the bowels, and apply cold water to the eye several times a day. If much inflammation exists, bleed from the small vein just below the eye, the course of which vein in all thin-skinned animals is quite distinctly marked.

SIMPLE OPHTHALMIA.

This disease arises sometimes from a blow inflicted by a passionate groom, or from some other external injury, or from a foreign body entering the eye, causing such an irritation in that delicate organ as sometimes to terminate in blindness.

The symptoms are considerable swelling and inflammation of the eye lids, their under surfaces being very much reddened, and the vessels highly injected with blood; there is also a cloudy appearance over the cornea, or transparent part of the eye.

For treatment, if the animal is in a plethoric condition, take six or eight quarts of blood from the jugular vein, regulating the quantity by the action on the pulse; otherwise general bleeding should not be undertaken. The bowels should be freely opened with Barbadoes aloes, six drachms; pulverized ginger root, one drachm; and pulverized gentian root, one drachm, made into a ball. Bathe the eye freely with cold water; after which apply with a syringe either of the following washes: laudanum, six drachms; rain, or distilled water, one pint; mix the two, and shake well before using:—or, take half an ounce of the extract of belladonna dissolved in one pint of rain water. Give internally one drachm of powdered colchicum morning and evening, in a bran mash; no grain should be given during the treatment; corn should be especially avoided.

SPECIFIC OPHTHALMIA.

Inflammation of the eye, or specific ophthalmia, is known to horsemen as moon-blindness, from the influence which the moon is supposed to exert upon it. This, however, is one of the many popular delusions which fill the pages of many useless works on farriery. When a horse is once attacked with this disease, he is ever after liable to subsequent attacks, at intervals varying from one to six months, and generally terminating in blindness. This termination may, however, be warded off for a long time by proper management; each subsequent attack rendering such a termination more and more certain, from the increased alteration in the structures of the eye.

The horse may appear perfectly well, and the eyes clear and bright one day, and the next morning usually one eye will be found closed, more particularly if it is exposed to a strong light; little or no swelling will be observed; the lining membrane of the eye lid is quite red, and the eye exceedingly watery and tender.

The causes of this disease are mainly attributable to hereditary predisposition, or to confinement in dark stables, and sudden exposure to strong light. Badly ventilated stables, in consequence of which the eyes are continually exposed to the strong fumes of ammonia arising from the urine, as also hard work in a small collar, are supposed to be exciting causes.

These cases require prompt attention, in order to ward off the serious consequences which otherwise are in store for the unfortunate animal. The bowels should first be opened with the purging ball recommended in simple ophthalmia. Give bran mashes only, and when the bowels are opened, give one of the following powders night and morning on the tongue :— pulverized colchicum, one and a half ounces; saltpetre, two ounces; divide into twelve powders. These will last one week, and by that time the eye will usually become clear and bright. Use as an injection for the eye, tincture of opium, one ounce; rain, or distilled water, one pint :—or, if more convenient, mix half an ounce of the extract of belladonna in one pint of water, and use in the same manner. If the animal is in a plethoric condition, bleeding will be found advantageous; the quantity to be regulated by the condition of the pulse. Place the animal in a cool, well-ventilated location, free from any ammoniacal gases.

CATARACT.

This disease, which is one of the terminations of specific ophthalmia, is an opacity of the crystalline lens, situated directly behind the pupil, through which it is visible. The first indications of cataract noticed are one or more white spots making their appearance within the eye, gradually enlarging, and at last blending with each other until the animal becomes totally blind. Not much can be done in such cases by way of treatment. Its removal by the operation practised upon the human eye, and known as "couching," is hardly advisable, as the horse is forever after unsafe, being very apt to shy at almost every object which he encounters, in consequence of his sight being but partially restored by the operation.

WALL EYE.

This peculiar appearance of the iris in some horses is not the result of disease, but is occasioned by the absence of what is called the pigment, which gives color to the eye. This pigment is secreted upon the inside of the iris, and where it does not exist, the iris, or that part of the eye which surrounds the pupil (so called from its brilliancy) remains white. Percival says: "It is a remarkable fact that this variety of hue in the iris corresponds with the color of the hair; bay and chestnut horses have hazel eyes; brown horses have brownish eyes; and very dark brown or black horses, eyes of a still darker, dusky brown shade. This curious relation is still more observable in human beings; the diversity of colors and hues in their irides being infinitely greater than any thing we behold among any one species of animals. Cream-colored and milk-white horses have wall eyes, and albinos have red eyes; in both which instances the iris is said to be destitute of any coloring matter whatever."

MISCELLANEOUS DISEASES.

POLL EVIL.

This disease arises from blows inflicted upon the poll, or back part of the head, of animals whose blood is impure, or in a morbid condition. Horses going in or out of stables with low doorways frequently strike their polls; pulling back upon the halter, and blows inflicted by passionate grooms, are among the exciting causes of this much dreaded complaint. The same injuries inflicted upon an animal in perfect health seldom cause any essential trouble; but when the blood is in a morbid condition, fistulous abscesses are formed, which are seldom curable by merely local treatment, even when the disease is treated in its earliest stages.

The author has no faith in the seton, so highly recommended in such cases, but depends principally upon constitutional treatment, which consists in first changing the condition of the blood from an unhealthy standard to a healthy one. This may be done by the proper use of alterative medicines, given in either of the following forms: Socotrine aloes pulverized, four ounces; soft soap, four ounces; linseed meal, one and a half pounds; mix with molasses so as to form a mass; dose, one ounce twice a day:—or, Socotrine aloes, eight ounces; soft soap, eight ounces; linseed meal, one and a half pounds; mix and dose as before:—or, the following powder may be used: sublimed sulphur, two pounds; sesqui-sulphuret of antimony in powder, one pound: dose, a tablespoonful twice a day in the feed. The sesqui-sulphuret of antimony should never be purchased in a powdered form, as it is often adulterated with lead, arsenic, magnesia, or iron; but should always be procured in conical masses.

THE RUNNING STALLION AMERICAN ECLIPSE.

If the abscess is soft and pointing, it should be opened, and a solution of zinc, two drachms to a quart of water, injected into the opening once or twice a day. A saturated solution of corrosive sublimate is sometimes used advantageously; though the zinc is much safer in the hands of inexperienced persons. The nux vomica, in half-drachm doses, is also used as an internal remedy with good effect.

FISTULA OF THE WITHERS.

This is precisely the same as poll evil, its location alone giving it a different name, and requires the same course of treatment. Its location is upon the raised part along the back, and over the shoulders, known as the withers, and it is caused by bruises from the forepart of the saddle, and other causes.

MELANOTIC TUMORS.

Swellings are generally termed tumors; but tumors proper are swellings in any part of the animal not attended by inflammation, comprehending bony, fatty, fibrous, melanotic, etc. For their removal an operation is generally requisite, which should be left to the veterinary surgeon. Melanotic, or black tumors are, however, peculiar to gray horses, and operations upon this class would be of little use, as the entire system is usually filled with them; where one is seen, many more may be suspected. The author presented to the Boston Veterinary Institute, some years ago, a large cluster, fifty or sixty in number, taken from the abdominal cavity of a gray horse, all united together by membranous attachments; and there could not have been less than ten thousand of these tumors in the animal from which this specimen was taken. They will frequently be found about the tail of gray horses, not being found in horses of any other color.

GLANDERS.

This fatal and much dreaded disease has baffled the efforts of veterinary surgeons in all ages of the world, and still continues so to do. It is decidedly contagious; yet as different diseases are often confounded with it, which may be detected by the competent practitioner, no animal should be condemned until the symptoms peculiar to glanders, which cannot well be mistaken if the disease is fully developed, have manifested themselves. The suspected animal should be removed and kept from all possible contact with any others. The author has deposited in the museum of the Veterinary College of Philadelphia the heads of a number of horses that were killed as glandered animals, and yet not one of them was so; the suspicious symptoms in each case arising from carious teeth. Animals afflicted with ozena have also frequently been killed as glandered; and in one case which recently came under the author's notice, where the animal was killed as glandered, the cause of all the difficulty was the filling up of the frontal sinuses by bony deposits.

It is necessary for the attendant to use the utmost caution when about a glandered horse, as the disease is freely communicated from the animal to man by inoculation. Of some sixty-seven cases reported in the Veterinarian of London as occurring in man, but three recovered, notwithstanding the utmost exertions of the ablest physicians that could be procured.

The most common cause of this disease is the impure air of close, ill-ventilated, and filthy stables, which acts injuriously upon the organs of respiration, destroys the constitution, debilitates the system, and renders it susceptible to the attacks of disease. Neglected catarrh, also, sometimes terminates in glanders; hard work and bad provender, together with sudden changes from exposure to cold and wet weather to hot stables, are likewise reckoned among the causes.

The symptoms are, discharges from one or both nostrils, of a glossy, thick, gluey nature, frequently sticking about the nostrils in considerable masses. This is a peculiarity which other discharges do not possess. This discharge is not always copious, as is generally supposed. The Schneiderian membrane of the nose changes to a dusky, or dirty yellow, or leaden hue; ulcers appear upon the membrane; a peculiar raising of the nasal bones will be observed, which the author has never noticed in any other disease; the discharge is sometimes mixed with blood, and is often fetid; and one or both of the submaxillary glands are swollen and adhere to the jaw bone. Too much reliance, however, should not be placed upon this swelling, as it frequently accompanies other diseases; but the character of the discharge, and the raising of the nasal bones are peculiarities not easily mistaken when the disease is developed. As all the other symptoms will be found accompanying other diseases, too much care cannot be exercised in deciding upon a case of this disease previous to a full development of the symptoms.

All treatment thus far has proven a failure.

FARCY.

This is regarded by the author as an incipient stage of glanders, or as a type of the same disease, and with proper management is curable. Experiments prove that the virus from a farcied horse will produce glanders by inoculation in a sound one, and that the glandered matter will in like manner produce farcy. There are two distinct varieties or stages of farcy: one, which is called button farcy, is altogether superficial, being confined to the lymphatic vessels of the skin, and readily yields to medical treatment; the other variety makes its appearance in the extremities, generally upon the inside of the hind legs, which become completely engorged; but the swelling is very different from the ligamentary thickening, or from œdema, being very uneven or lumpy, excessively tender, and painful to the touch. Small abscesses are formed, which at first discharge a healthy pus, but soon ulcerate, and discharge a thin, sanious matter. These abscesses, or tumors, first make their appearance on the inside of the hind legs, and then on the fore ones in like manner; the neck and lips come next in turn, and they may afterward appear in all parts of the body, when glanders will begin to manifest itself.

By way of treatment, good wholesome food is all important. Sulphate of copper in two-drachm doses, combined with one or two drachms of pulverized gentian root, will often prove successful; corrosive sublimate, also, in ten or fifteen grain doses, night and morning, has often been advantageously used; the doses may be increased to a scruple, or even half a drachm, if the animal bears the medicine well. If the animal is much debilitated, give calomel in half-drachm doses instead of the sublimate, or the sulphuret of mercury may be substituted. The use of arsenic has also been attended with good success, but the author has been more fortunate with the muriate of baryta in half-drachm doses, than with any other preparation in use. All the tumors should be opened, and caustic carefully applied to each; sulphate of copper, nitrate of silver, the per-manganate of potash, or the red-hot iron, are the best applications. The following ointment should be rubbed along the corded vessels once a day: blue ointment, two ounces; hydriodate of potash, two drachms; lard, two ounces; mix well. Or, the red oxide of mercury, two drachms to the ounce of lard is very good.

SCARLET FEVER.

This disease, otherwise called scarlatina, has too frequently been confounded with farcy, notwithstanding the two diseases present very different symptoms. It is easily managed, yielding readily to medical treatment.

For several days previous to any very noticeable symptoms being manifest, the animal is off his feed, dull and mopy, with mouth hot and dry; slight or copious discharges from the nose, mixed with blood; the Schneiderian membrane highly colored, and presenting numerous scarlet blotches, irregular in form, and containing a thin, reddish fluid; these blotches sometimes present a pustular appearance, but upon passing the finger over them, nothing of that character can be discovered; the whole body is covered with similar spots, which sometimes require close examination to discover them; in other cases, little pustules are formed, which break, and discharge a thin sanious fluid of a reddish color and glucy nature; swellings of the legs, sheath, and belly, are usually attending symptoms; the respiration is quick; the pulse is full and accelerated, and there is a disinclination to move.

For treatment, the extract of belladonna alone appears to be a specific in this disease. It should be given in doses of from one half to two drachms, dissolved in a pail of water, and given to the animal to drink. No hay should be placed before him; soft mashes only should be allowed, until he is convalescent. This treatment, so simple yet so effective, has been pursued by the author's friend, Dr. Bowler, of Cincinnati, and himself, for the last ten years, with uniform success, not a single case having been lost. It is true that the disease is not of very common occurrence; yet during that period the author has had over thirty cases.

MANGE.

Diseases of the skin are less numerous in the horse than perhaps, most other animals; a circumstance doubtless arising from the great care taken of our better class of horses to keep the skin clean, thereby promoting its healthy action. Mange is identical with the itch in the human body, and is an infectious disease, the intolerable itching being caused by minute insects, called *acari*. They are first observed with the aid of a powerful microscope along the mane and the root of the tail, causing a scurfy appearance of the skin. This appearance rapidly extends to the neck and body; spots denuded of hair will appear, which gradually run into continuous scabby patches. As the disease advances, it thickens and puckers the skin, particularly of the neck, withers and loins. This disease is easily cured if properly managed.

The natural history of these insects is not well known. They live only upon, or beneath, the skin of animals. There appears to be a distinct variety, peculiar to each species of animals. They live for a considerable time after being removed from the skin, but for how long a period is not precisely known. According to experiments made it appears that they can live in pure water for three hours; in strong vinegar, alcohol, and in a solution of kali carbonicum, twenty minutes; in a solution of sulphuric acid, twelve minutes; in turpentine, nine minutes; and in a solution of arsenic, four minutes.

THE THREE FRIENDS.

When they are not exposed to such violent and destructive immersions, it has been said that they will retain life for six or eight days. From a comparison of the *acari* of mangy animals, it is supposed that the variety peculiar to the horse can live for a much longer period. Mangy horses have been removed from their stalls, washed with various preparations, put into another stable, and completely freed from the effects of the disease; but upon returning to their former stalls, or using unwashed their accustomed harness, the disease soon showed symptoms of its return. This fact accounts for the trouble experienced in curing this disease. The insect is rubbed off upon the sides of the stall, or clings to the harness, again to come in contact with the animal.

For treatment, the animal should be stripped of all harness, well washed with acetic acid, and turned into a loose box stall away from that in which he has been standing. If this course is adopted, one or two washings will generally suffice. The harness also should be well washed, and not used for two or three months; nor should the horse be replaced in his former stall for a less period, and not even then until it has been thoroughly cleaned and white washed. A wash of white hellebore and water has been much and beneficially used for this disease; and in inveterate cases corrosive sublimate in solution is recommended, though there is some danger of its absorption; if this should occur, the animal would quite likely be destroyed. A mixture of sulphur, oil, and turpentine is highly recommended; but the author has never witnessed the superior qualities of any of these preparations over the acetic acid.

SURFEIT.

This disease appears all over the body in the form of pustules, which seem scaly, and then appear to get entirely well, while fresh ones make their appearance, and follow in the same course. The hair is rough, staring, and unhealthy in appearance; the legs sometimes become much swollen, and there is general debility. This disease is supposed to arise from bad grooming, bad management, and unwholesome food, together with a general plethoric state of the system.

For treatment, bleed the animal if plethoric, taking from

the neck vein from four to six quarts; in the absence of plethora, the lancet must not be used. Give a strong purging ball, followed by one of these powders twice a day: saltpetre, one and a half ounces; flower of sulphur, two ounces; black antimony, one ounce; mix and divide into eight powders.

HIDE BOUND.

This is a condition of the skin, caused by some morbid action in the system. Derangement of the digestive organs will induce it. The animal must be treated for the disease under which it is laboring.

STRAINS OF THE LOINS.

Strains are of very frequent occurrence in the horse, in consequence, doubtless, of the great amount of labor demanded of him, which often taxes his powers to the utmost. These strains frequently give rise to serious trouble, rendering the animal unfit for work and often establishing an incurable lameness. Strains of the loins occur most frequently in draft horses, particularly in those used in the shafts of drays or carts. Such animals on going down hill heavily loaded are very apt to become injured; at times the injury is so great that the spinal marrow becomes affected, causing paralysis of the hind extremities, and rendering the animal comparatively useless ever after. When the injury is very severe, bleeding should be resorted to, if the animal can bear it. The following liniment will be found an excellent application for strains of all kinds: laudanum, gum camphor, spirits of turpentine, tincture of myrrh, **castile soap,** oil origanum, nitrous ether. of each one ounce;

alcohol, one quart; mix all together, and shake well before using; apply two or three times as occasion may require.

PALSY.

This is a loss of power in the nervous system. General palsy is never found in the horse, it being always partial or limited in extent, and described under two heads, paraplegia and hemiplegia. The first is a paralysis of the hind extremities, which is of very frequent occurrence; it sometimes occurs as a sympathetic affection, in which cases it disappears with the other symptoms of the disease. The second form is a palsy of one side of the body only, and is of very rare occurrence. When paralysis arises from strains whereby the spinal cord is injured, it causes the most acute suffering, and the animal usually dies in a few days. When the pressure upon the spinal cord is not great, the animal is sometimes rendered useful for ordinary purposes, but very rarely becomes sound.

For treatment, first open the bowels if they are the least costive, and give internally one of the following powders night and morning; nux vomica, one ounce; pulverized gentian root, two ounces; Jamaica ginger, one ounce; mix, and divide into twelve powders. Apply warm sheep-skins to the loins, succeeded by the following application: linseed oil, one pint; spirits of hartshorn, four ounces; shake well before using. Perfect rest and moderate diet are necessary.

LOCKED JAW.

This distressing malady, otherwise known as tetanus or trismus, is one generally arising from neglected wounds, such as are occasioned by a horse's picking up a nail; in which case the wound, instead of being kept open by the owner, or his attendant, is suffered to close up, in consequence of which, if there is the slightest disposition to ulceration, matter is formed under the horn or hoof, which develops the most alarming symptoms, usually in about two weeks after the wound has healed. When locked jaw is the result of wounds, it is called symptomatic, or traumatic; when existing without apparent cause, it is called idiopathic. The latter is said to be caused in some cases by the action of bots and of worms in the intestines.

The first symptoms observable are a stiff, straggling gait behind; rigidity of the muscles of the jaw, completely locking the jaws together; the tongue is sometimes swollen, and considerable saliva flows from the mouth. As the disease progresses, the muscles throughout the body become rigid; the animal turns as though there was not a joint in the body; the nose is poked out, the nostrils dilated, and respiration disturbed; the bowels are almost invariably constipated; on elevating the head, a spasmodic or flickering motion of the eye will be observed, exposing little more than the white parts. When the disease is confined to the head and neck, it is called *trismus;* when extended to all parts of the body, it is termed *tetanus.*

There can scarcely be any principle laid down to govern the treatment of this disease, as cases have recovered under all kinds of treatment. The great object is to get the bowels opened; when this is accomplished, the cases usually have a favorable termination; but when the jaws are firmly set, the prospects are very limited. Give, if possible, by the mouth one ounce of aloes, ten drops of croton oil, two drachms of pulverized gentian root, and one drachm of ginger; make into one ball with molasses.

If this cannot be given, keep a ball of aloes in the mouth, the action of which may be increased by adding to the ball two drachms of calomel, and omitting the croton oil. Give injections of belladonna, half an ounce dissolved in a pail of water. Opium has been much used, but is giving way to other preparations. Give upon

BYRON'S MAZEPPA.
"They left me there to my despair,
Link'd to the dead and stiffening wretch."

the tongue every hour twenty drops of the following mixture: hydrocyanic acid and tincture of aconite, of each one ounce; mix, and shake, well together. Blistering the back, from the head to the tail, has succeeded in some cases. Chloroform has been highly recommended, but appears to have only a temporary effect; it is given in doses of from one to two drachms.

RHEUMATISM.

This disease is quite common in the Western States. The symptoms are stiffness, lameness, and shifting from one limb to another; sometimes tumefaction is observable about the extremities. The lameness is sometimes absent, and appears to be influenced by changes in the weather.

For treatment, poultice the feet with mustard and flaxseed meal. Give internally of nux vomica, one ounce; pulverized gentian root, one and a half ounces; pulverized ginger, one ounce; mix, and divide into twelve powders; give one every night in the feed.

The most successful treatment which the author has found, when the above has failed to effect a perfect cure, is that recommended by Dr. Bowler, of Cincinnati, whose experience in baffling this disease has been quite considerable. It is as follows:—if the animal is plethoric, bleed freely and give a strong cathartic; follow every morning with one of the following balls: pine tar, two ounces; pulverized gentian root, one ounce; mix all together, and divide into eight balls. Keep the body warm, and give no corn.

CRAMP.

This complaint occasions considerable alarm to the owner of a horse, from the peculiarity of the symptoms. A horse is found to go suddenly lame, lameness continuing, dragging one leg after him as though it were dislocated or broken. Upon taking a whip and striking him, he will sometimes go two or three steps in a natural way, and then the leg drags again. Such instances have been pronounced fractures by the farrier, and even by the young veterinarian such a mistake has been made; indeed, there are instances of the horse's having been killed by order of the medical attendant.

Treatment. Friction by hand-rubbing, and application of the liniment recommended for strains. Usually the animal will be found all right upon the following day.

HYDROCELE.

This disease commonly known as dropsy of the testicles, sometimes affects the stallion. It consists of a collection of serum in the *tunica vaginalis*, or bag containing the testes, fluctuating when pressed by the hand, but free from tenderness or pain. Its causes are obscure, but it is supposed to result from injuries, such as strains, etc.

For treatment, the scrotum should be punctured, and a weak solution of tincture of iodine injected into the *tunica vaginalis;* or equal parts of port wine, and water of zinc lotion, or lime water, may be used with very good effect. The animal should be well secured before these preparations, particularly the first, are used, as the pain thereby caused may render him for the time unmanageable.

WARTS.

These fungous growths appear in the horse most frequently about the mouth, nose, and lips; but they are occasionally found upon other parts of the body. They are sometimes found in large numbers about the lips of colts, and are generally rubbed off, or drop off; if, however, they grow large and become deeply rooted, they may be cut off by passing a needle through the centre armed with double thread, and tied tightly around the neck on each side. This prevents the possibility of the ligatures being rubbed off. Or, they may be painted over with the per-manganate of potash, a few applications of which will entirely destroy warts of a large size; or they may be removed with a knife.

SIT-FASTS.

These are dark, hard, scabby spots upon the back, which are dead skin and cannot be easily removed; but by poulticing for several days they become soft and may be torn off. Tincture of myrrh applied two or three times a day will generally effect a cure after the dead skin is removed.

WARBLES.

These arise from bruises, which cause superficial swellings that sometimes suppurate. They should be freely opened and the matter well washed out. A solution of sulphate of zinc, or alum-water, is all that is required to effect a cure.

SADDLE OR HARNESS GALLS.

These are bruises caused by friction and moisture, occurring most frequently in warm weather; the parts are rubbed raw, and sometimes bleed. The treatment is simple and effectual. Bathe the parts several times a day with one pint of water and half a pint of tincture of myrrh.

MALLANDERS AND SELLENDERS.

These are scurfy eruptions of the back part of the knee joint and the front part of the hock joint. They sometimes occasion much pain, and lameness in consequence. They constitute but one disease, the names having reference to the fore and hind extremities; mallenders being applied to eruptions upon the fore extremities, and sellenders to those upon the hind ones.

For treatment, wash the parts well with castile soap and water, and apply the following: lard, four ounces, and Goulard's extract, one ounce, well mixed.

ULCERATION OF THE UDDER.

Mares are sometimes subject to this disease, which is caused by the milk's coagulating in the bag, and causing inflammation and suppuration. The udder becomes swollen, hot, tender, hard, and knotty. A flaxseed poultice should at once be applied, when the abscess will soon be brought to a head, which will be known by its smooth, polished appearance and its soft feeling. It should then be lanced, and the udder bathed twice a day with lard melted as hot as the animal can bear. Sometimes it becomes necessary to inject a solution of the sulphate of zinc into the opening; but in ordinary cases the hot lard is sufficient, if properly applied.

INFLAMED VEINS.

The jugular or neck vein sometimes becomes inflamed in consequence of being injured by a bungling bleeder. A swelling is first noticed, followed by a gaping in the incision in the neck, from which an acrid fluid oozes.

For treatment, bathe the part well with cold water, into which a small portion of tincture of myrrh is thrown, and with a purging ball a cure is soon effected.

SURGICAL CASES.

It frequently becomes necessary, in order to relieve the animal from some painful disease, to resort to operations in surgery; this, in fact, has of late years become an important branch of veterinary practice. When it becomes necessary to use the knife, the animal should be spared all useless torture. In severe operations, humanity dictates the use of some anæsthetic agent to render the animal insensible to pain. Chloroform is the most powerful of this class, and may be administered with perfect safety, provided a moderate quantity of atmospheric air is inhaled with or during its administration. Sulphuric ether acts very feebly upon the horse, and cannot therefore be successfully used. Chloric ether answers a very good purpose, but pure chloroform is preferable. In minor operations, the twitch, the side-hobble, or the foot-strap, is all that is necessary. When a horse is to be cast for an operation, force must be used for its accomplishment. The patent hobbles have been preferred for that purpose by veterinary surgeons generally, though the author prefers a modification of the cast-rope and the patent hobbles. This improvement consists in having a heavy, well-padded leathern collar, each layer burned in with rosin, after the style of the old-fashioned fire-buckets; at the bottom of this collar a strong ring is attached, secured by an iron band; through this ring the rope is passed; around the body a strong leathern band is buckled, which connects with the top of the collar by a cross strap, which keeps it in place; a hobble band is placed upon each hind fetlock, through the D of which the rope is passed; on each side of the collar a strong ring is firmly secured, through which the rope also passes, the ends of which are then pulled upon by one or two men on each side, and the animal let quietly down. The author is convinced by experience that this arrangement is far preferable to any hobble arrangement yet seen. It is a mistaken idea that horses must be cast for **every little operation**; in truth, but few operations require it.

BLEEDING.

Blood-letting in former times was regarded as the sheet-anchor in veterinary practice; but that day has past. The practice of bleeding horses upon all occasions cannot be too strongly condemned; the cases where blood-letting proves beneficial being comparatively few. Before using the lancet the pulse must be examined, the condition of the animal considered, and the effects upon that pulse must decide the quantity of blood to be taken. The pulse will be found following the front margin of the masseter muscle, which muscle forms the fleshy parts of the head upon each side, called the cheeks. By following the front part of this muscle downward with the thumb, until near the base of the lower jaw, and then passing the forefingers under, or inside of the jaw, the pulse will be readily felt; or, to point its location out with more certainty, if an imaginary line is drawn perpendicularly from the front part of the ear downward, it will cross the point where the pulse is located and felt.

LADY SUFFOLK.

In a healthy condition the pulse beats from thirty-six to forty times a minute; variation above or below this standard indicates a morbid condition of the system. This fact should be born in mind in the description of any disease. When bleeding is necessary, the neck never should be corded, as much injury has at times been caused by this practice. All that is requisite is to raise the jugular vein by pressing upon it with the fingers of the left hand, using the lance with the right. The old-fashioned mode of bleeding with the fleam and blood-stick is a bungling operation, frequently requiring several trials before bringing blood, the result of which is an inflamed vein. A more convenient, a more certain, and a more satisfactory method is by using a spring lance, made for the purpose, which never fails in bringing blood upon the first trial. It is so contrived as to straddle the vein of the neck, which keeps it firm, and prevents its rolling, so that it is impossible to miss bringing the blood when it is once placed upon the vein and sprung. By this method of bleeding, the covering of the eye and the cording of the neck are unnecessary, and the operation can easily be accomplished by one person. After the vein has been opened, the blood-pail pressed against the vein will cause the blood to flow freely. When the desired quantity has been drawn, the vein must be carefully closed by passing a pin through the centre of the opening, taking up the skin upon both sides, and tying with hair from the mane or tail. The pin may be removed in about twenty-four hours.

NEUROTOMY OR NERVING.

This is one of the most important operations in veterinary practice, and one that has been much abused, not only in Europe, but even more so in the United States. Its usefulness was first demonstrated by Assistant Professor Sewell, of the Veterinary College of London. The operation consists in cutting out a portion of the metacarpal nerves on each side of the legs, thus destroying the sensibility of the foot. From the instantaneous relief experienced by the animal in all cases of foot lameness, no matter from what cause, an opportunity has been afforded to dishonest persons for imposing upon the public by availing themselves of this practice; an opportunity, it need not be said, which has been freely used, and thus a valuable operation has been brought into undeserved disrepute. The cases likely to be benefited by this operation are few, and should be selected with great care; otherwise the loss of the animal's hoof may be, and often is, the termination of the case.

This operation is recommended by veterinary authors in incurable cases of lameness of the navicular joint; but sufficient caution is not impressed upon the mind of the reader, to enable him to guard against the fatal results which too often follow.

In deciding upon a case for this operation, an animal should be selected with a foot as free from contraction as possible; free from corns; comparatively free from inflammation; with a concave ground surface; open heels; hoof free from rings or roughness; and no bony deposits within the hoof. In such a case, the operation may be performed with success. A horse that has been foundered should not, under any circumstances, be operated upon, as ossification of the laminæ frequently follows such an attack; nor a horse affected with ossification of the lateral cartilages, corns, or badly contracted hoof; for these are the cases where loss of the hoof is likely to follow, rendering the animal useless.

After the operation has been performed, care should be taken in driving the animal; for it should be remembered that no matter what accident may happen to the foot, the animal is unconscious of pain. The feet should be frequently examined to see whether the horse has picked up a nail, or otherwise injured the foot; for such injuries would otherwise remain undiscovered until too late to save the animal's life or usefulness. The smith should be informed of the operation, in order to guard against pricking the animal's foot in shoeing.

It is necessary previous to the operation that the feet should be perfectly cool, which condition may be obtained by frequent bathings with cold water for several days previous. The horse is cast, the foot to be operated upon loosened, and brought forward by an assistant, it resting upon a bed of straw. A vertical incision is made about two inches above the fetlock, between the cannon bone and back sinew, raising up with the forceps the cellular membrane, and carefully dissecting out the nerve. The precaution should be taken of placing the finger upon it, as the artery has been taken up and cut off before the mistake was discovered. Having fairly exposed the nerve, pass a curved needle armed with strong thread under it, and by carefully drawing it up and down the nerve may be readily separated. A sheathed knife is then passed under the nerve, and by a quick motion the nerve is severed at the upper part. After the struggles of the animal cease, the cut nerve may be raised with the forceps, and from one-half of an inch to an inch removed. This second cut causes no pain. The wound is then closed by three single stitches. After operating upon both sides in like manner, the animal is allowed to rise. Bandages should then be placed upon the leg, and kept saturated for several days with cold water.

LITHOTOMY.

Operations for stone in the bladder of the horse have been practised since 1774, and in many cases very successfully. In performing this operation, an ordinary scalpel, a probe-pointed bistoury, a fluted whalebone staff, and a pair of curved forceps are necessary. The animal should be placed upon his back with the hind legs drawn well forward; a whalebone staff is passed up the urethra, which may be felt a little below the anus; an incision, one and a half or two inches in length is made directly upon it, obliquely to one side, cutting through the urethra and the neck of the bladder; the forceps are next introduced, and the stone removed; after which the parts are carefully closed by means of the quill suture, which in this operation is far superior to the interrupted one, as it more effectually prevents the dribbling of urine through the wound, which always occurs with the interrupted one, and therefore causes a more speedy union of the parts.

TREPHINING.

This operation consists in cutting out circular pieces of bone with a circular saw, called a trephine, and is most generally performed in cases of fracture of the skull, or face. The bone removed must be from the sound part contiguous to the fracture, so as to enable an elevator to be passed inside of the cranial case, for the purpose of pushing back the broken bone to its proper place, and removing all detached pieces. This operation is also performed in cases of ozena, by removing a piece of bone over the frontal sinuses, situated immediately between the eyes, in order to expose the diseased parts at once, that they may be washed with proper injections.

TENOTOMY.

This operation is practised for the purpose of strengthening crooked legs or sprung knees. It consists in dividing the flexor tendons, in order to bring the limb straight. There are but few cases, however, in which the operation would be of much service, and therefore care must be exercised in selecting such cases as are proper. It would hardly be proper in a young horse, as other means less objectionable often succeed. In old horses it would not be prudent, as their limbs are generally stiff and permanently set; nor would it be successful in cases where anchylosis or stiff joint existed, as is often found in connection with crooked legs and sprung knees.

COUCHING.

This is an operation upon the eye for the purpose of removing a cataract from the axis of vision. A couching needle is passed through the sclerotic coat of the eye a little behind the cornea, passing it upward behind the iris to where the cataract is located, pressing it downward into the vitreous humor behind the iris, where it remains. This operation has not been very successful in the horse, by reason of the imperfect restoration of the sight thereby afforded, which causes them in almost every instance to shy at every object which they encounter, thus rendering them dangerous upon the road.

TAPPING THE CHEST.

This operation consists in passing a round, pointed instrument, sheathed with a canula, into the chest, in order to draw off any accumulation of fluid that may have taken place in the viscus. The instrument is passed, after first making a small incision through the skin, between the eighth and ninth ribs, but not too low down. It is pushed gently forward until it penetrates the pleura, or lining membrane of the chest. The stellet is then withdrawn, and the canula is kept in place until the fluid ceases to run. If, however, a large quantity exists, all of it should not be taken away at one time; for the pressure upon the lungs having been so great, if such sudden relief is afforded, nature, unable to accommodate herself to so rapid an alteration gives way, and the animal consequently dies. It should therefore be taken away at one, two, or three tappings, as occasion may require. Good wholesome food should be allowed.

PERIOSTEOTOMY.

This operation is most generally performed for painful splints. It consists in cutting though the periosteum, or membrane covering the surface of all bones, over the splint or node, which immediately gives relief. This operation requires the aid of an experienced man.

AMPUTATION OF THE PENIS.

This operation is occasionally called for in the horse, particularly in cases of paraphymosis, or protrusion of the penis, that have resisted all other modes of treatment. The operation, as performed in England, is unnecessarily tedious, and not as successful as it should be. It is only requisite in performing this operation to place a twitch upon the animal, and while he is standing to take the penis in the left hand, and with an amputating knife in the right to sever it at one stroke. The hemorrhage, although considerable, need not occasion any alarm. A piece of cotton or soft sponge, saturated with spirits of turpentine or any other styptic, and placed in the sheath, will soon cause the hemorrhage to cease. Fear of hemorrhage, may deter some persons from performing what may appear a bold operation; but the author has not known a single operation performed in this way to have a fatal termination; whereas with the English mode of operating it frequently does so, beside, even if it is successful, rendering the animal useless for a much greater period of time.

ŒSOPHAGOTOMY.

This operation is occasionally resorted to where any foreign substance, as an apple, potato, carrot, and the like, has lodged in the œsophagus, or gullet. Where such obstructions exist, gentle manipulations with the hand should first be resorted to; if these are not successful in removing them, the probang is called for, and in case of failure thus to dislodge them, this operation is the only remaining resort. It is not necessary to cast the animal. Cut down directly upon the swollen part of the throat, and remove the obstruction. The wound may then be closed by means of the interrupted suture; that is, by single stitches, at proper distances apart, allowing the ends to hang out of the external wound, which may be closed in the same manner. The animal should be kept on gruel for several days. If the gruel is seen to ooze out of the wound when he is swallowing, it should be carefully washed away with cold water. The parts should be syringed with a weak solution of sulphate of zinc, chloride of zinc, or tincture of myrrh.

HERNIA.

By the term hernia surgeons understand a rupture, or protrusion of some of the viscera out of the abdomen, forming a soft tumor. In human practice there are hernias occurring in all the viscera of the body; but in the equine race they are confined, with rare exceptions, to the abdominal viscera, the inguinal hernia being the most common. This appears in the groin, and is a protrusion of the intestine through the abdominal ring, which in the stallion frequently passes down into the scrotum, or bag, constituting scrotal hernia. These hernias sometimes occur during castration in consequence of the violent struggles of the animal. In such cases it is best to administer chloroform at once in order to quiet the animal and prevent violent strugglings. The animal should be put upon his back, and one hand passed up the rectum, and one or two fingers of the other placed upon the scrotum, when by careful manipulations the intestine can generally be replaced. If, however, a reduction cannot be effected, an operation will be necessary. The hernia should be exposed by cutting through the integument a little upon one side, and coming down upon the hernia, the finger is placed upon it,

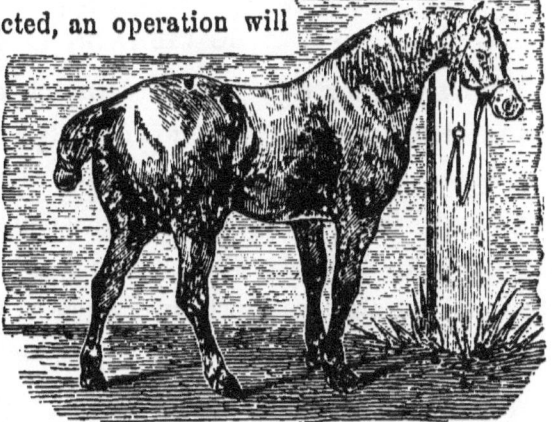

GOOD FOR HEAVY DRAFTS.

and a reduction effected by careful manipulation. Care should be taken that the nails upon the hand are trimmed close, in order to prevent wounding the intestine. The wound should then be closed by means of the interrupted suture. A folded cloth should then be applied to the part, and retained by means of a continuous bandage crossed between the legs from side to side in the form of the figure 8. Sometimes the intestine becomes strangulated, constituting strangulated hernia, the reduction of which requires an operation as before mentioned. If, however, it is found impossible, then to reduce it, the finger should be passed through the opening, if possible, and a probe-pointed bistoury following upon it, enlarge the opening and replace the intestine. The same treatment as before indicated will be necessary.

The symptoms of strangulated hernia are very similar to those of acute enteritis, or inflammation of the bowels. These may be regarded as the only hernias to which the horse is liable.

ROWELING.

Rowels were formerly much used, but of late years the seton has superseded them. The rowel consists of a round piece of sole leather, cut out in the centre wound round with tow, which is saturated before using with digestive ointment. The skin is cut through, and dissected upon each side sufficiently to admit the rowel. This is used principally under the jaws and in the breast. The seton answers the same purpose, and is much more convenient. It consists in arming a needle made for the purpose with tape and passing it through the part desired, the seton being saturated with the same ointment as the rowel.

FIRING.

The object in firing a horse is to produce an external inflammation where counter-action is required, as in spavin, ringbone, curbs, etc. The operation may be performed upon the animal while standing, by placing a twitch and side line upon him; but if the surface to be fired is extensive, and the animal high strung, it is better to cast him, particularly where a number of oblique, vertical, or horizontal lines are to be drawn. Firing is not practised at the present day to the extent that it formerly was, and when it is practised every endeavor should be made to prevent, as far as possible, the blemishes which always follow the operation. Various forms of irons have been adopted to accomplish this end. The author gives the preference to the feathered iron, which is brought down to a very fine edge, and, opinions are entertained by veterinary surgeons as to the advantages resulting from deep firing as compared with those accruing from surface firing. In the author's judgment, if firing is resorted to at all, it should be done effectually. His attention has recently been called to a firing iron devised by A. Maillard, Esq., of Bordentown, New Jersey, which is the best adapted instrument that has ever passed under his notice. It consists of two pieces of iron, octagonal in form, about one and a half inches long by one and a quarter wide, one piece containing five round-pointed projections, placed one at each corner and one at the centre, and the other four points, so arranged as, when fitted together, to fill up the intermediate spaces of its opposite; both irons being used alternately on the same parts without extending the surface fired. This iron will probably supersede any iron in use, and thanks are due to the inventor for his ingenuity in producing it. Pointed instruments have been before used, but far inferior in their arrangement.

TRACHEOTOMY.

This operation is occasionally called for in cases of strangles, when the swelling threatens suffocation, as it is often the only means of saving the animal's life. It consists in making a longitudinal incision through the skin immediately over the windpipe and below the larynx, cutting through the cartilaginous rings (two or more, as occasion requires), and inserting in the opening a tube of silver made for the purpose, through which the animal breathes, instead of through the nostrils. A circular piece is sometimes cut out of the windpipe in order to admit the tube more freely, which is certainly the better mode of performing the operation. In a case of emergency, a piece of elder with the pith pushed out will answer temporary purposes. It should be well secured from slipping into the windpipe by means of a piece of string.

THE HORSE TAMED.

RAREY'S METHOD OF TAMING HORSES.

THE great celebrity which Mr. Rarey obtained in England and France, owing to his unparalleled success in rendering the most vicious and ungovernable horses perfectly tractable and gentle, has excited no small degree of interest and curiosity among us, to ascertain the method which he adopts to secure such noteworthy results. To gratify this interest, as laudable as it is natural, we propose in this place giving the leading features of his method, as gleaned from the various English publications bearing upon the subject, especially from the little work, entitled "The Art of Taming Horses. By J. S. Rarey," and edited by the Hunting Correspondent of "The Illustrated London News."

It is needless to premise, that not every man can become a Rarey, by the perusal of this, or of any other treatise upon the art of breaking horses; yet it is not claiming too much for this system to say, that by its use the large majority of horses may be broken more expeditiously, more effectually, and with far more satisfaction and pleasure to the breaker than by the adoption of any other now known. It is no slight gain, to be able to transfer the breaking of horses from ignorant, impatient, and disagreeable persons to those who can in every respect appreciate the noble qualities of the animal and who will therefore deal with him as his high rank in the scale of creation demands.

The three fundamental principles of the Rarey theory are: first, that the horse is so constituted by nature that he will not offer resistance to any demand made of him which he fully comprehends, if such demand is made in a way consistent with the laws of his nature; second, that he has no consciousness of his strength beyond his experience, and can be handled according to our will without force; and third, that we can, in compliance with the laws of his nature, by which he examines all things new to him, take any object however frightful around, over, or on him, that does not inflict pain, without causing him to fear.

As to the first proposition:—the horse, although possessed of some faculties superior to man's yet being deficient in reasoning powers, has no knowledge of right or wrong, of free will and independent government, and is not aware of any imposition practised upon him, however unreasonable it may be. He cannot, consequently, decide as to what he should, or should not do, not having the requisite faculties to enable him to argue the justice of the thing demanded of him. Had he such faculties, taking into consideration his superior strength, he would be useless to man as a servant. If he had mind in proportion to his strength, he would roam through the fields at large, yielding service to no one. His nature has been wisely formed to be operated upon by the knowledge of man according to the dictates of his will, and he may properly be termed an unconscious submissive servant. This truth is verified in every day's experience by the abuse to which he is subjected. Any one who chooses to be so cruel can mount the noble steed, and run him till he drops with fatigue, or, as is often the case with the more spirited, falls dead beneath his rider. If he had the power to reason, would he not rear and pitch his rider, rather than suffer him to run him to death? Happily for us, he has no thought of disobedience, except by impulse caused by the violation of the law of his nature. If then, he is disobedient, it is the fault of man.

As to the second: the fact that the horse is unconscious of the amount of his strength, can be proven to the satisfaction of any one. Were it otherwise, the light vehicle in which he is placed, the slender reins and harness which guide and confine him, would be snapped asunder in an instant, at his own volition; no hitching-post could restrain him against his will, no saddle girth be placed around his body. These facts, made common by every-day occurrence, are not regarded as anything wonderful. Their continued existence serves to remove us from all consideration of them.

As to the third: there being, as we know from a natural course of reasoning, some cause for every impulse or movement of either mind or action, and this law governing every action or movement of the animal kingdom, there must be some cause before fear can exist; and if fear exists from the effects of imagination, and not from the infliction of real pain, it can be removed by complying with those laws of nature, by which the horse examines an object, and determines upon its innocence or harm.

A log or stump by the roadside, for example, may be, in the imagination of the horse, some great beast about to pounce upon him; but after he is taken up to it, and allowed to stand by it for a little time, and to touch it with his nose, and to go through his process of examination, he will not care anything more about it. The same principle and process will have the same effect with any other object, however frightful in appearance, in which there is no harm.

These principles being taken as the basis, whatever obstacles oppose the proper breaking of horses are readily surmounted by the Rarey method, commencing with the first steps to be taken with the colt, and thence proceeding through the whole task of breaking.

How to Call a Colt from Pasture.—Go to the pasture and walk around the whole herd quietly, and at such a distance as not to cause them to scare and run. Then approach them very slowly, and if they stick up their heads and seem to be frightened, stand still until they become quiet, so as not to make them run before you are close enough to drive them in the direction you want them to go. And when you begin to drive, do not flourish your arms or halloo, but gently follow them, leaving the direction open that you wish them to take. Thus taking advantage of their ignorance, you will be able to get them into the pound as easily as the hunter drives the quails into his net. For, if they have always run in the pasture uncared for (as many horses do in prairie countries and on large plantations), there is no reason why they should not be as wild as the sportman's birds, and require the same gentle treatment, if you want to get them without trouble; for the horse, in his natural state is as wild as a stag, or any of the undomesticated animals, though more easily tamed.

How to Stable a Colt without Trouble.—The next step will be, to get the horse into a stable or shed. This should be done as quietly as possible, so as not to excite any suspicion in the horse of any danger befalling him. The best way to do this is to lead a broken horse into the stable first, and hitch him, then quietly walk around the colt and let him go in of his own accord. This should be undertaken slowly and considerately, as one wrong move may frighten your horse, and make him think it necessary to escape at all hazards for the safety of his life—and thus make two hours' work of a ten minutes' job; and this would be all your own fault, and entirely unnecessary—*for he will not run unless you run after him, and that would not be good policy unless you knew that you could outrun him, for you will have to let him stop of his own accord after all.* But he will not try to break away unless you attempt to force him into measures. If he does not see the way at once, and is a little fretful about going in, do not undertake to drive him, but give him a little less room outside, by gently closing in around him. Do not raise your arms, but let them hang at your side, for you might as well raise a club: *the horse has never studied anatomy, and does not know but that they will unhinge themselves and fly at him.* If he attempts to turn back, walk before him, but do not run; if he gets past you, encircle him again in the same quiet manner, and he will soon find that you are not going to hurt him; and then you can walk so close around him that he will go into the stable for more room, and to get further from you. As soon as he is in, remove the quiet horse and shut the door. This will be his first notion of confinement—not knowing how he got into such a place, nor how to get out of it. That he may take it as quietly as possible, see that the shed is entirely free from dogs, chickens, or anything that would annoy him. Then give him a few ears of corn, and let him remain alone fifteen or twenty minutes, until he has examined his apartment, and become reconciled to his confinement.

While he is eating, see that your halter is ready and all right, and determine for yourself the best mode of operation. Always use a leather halter, and be sure to have it made so that it will not draw tight around his nose if he pulls on it. It should be of the right size to fit his head easily and nicely, so that the nose-band will not be too tight or too loose. Never put a rope halter on an unbroken colt, under any circumstances whatever. Rope halters have caused more horses to hurt or kill themselves than would pay for twice the cost of all the leather halters that have ever been used for the

BRIDLE WITH A WOODEN GAG-BIT FOR CONQUERING VICIOUS HORSES.

purpose of breaking colts. It is almost impossible to break a colt that is very wild with a rope halter, without having him pull, rear, and throw himself, and thus endanger his life; and this, because it is just as natural for a horse to try to get his head out of anything that hurts it, or feels unpleasant, as it would be for you to try to get your hand out of a fire. The cords of the rope are hard and cutting; this makes him raise his head and draw on it, and as soon as he pulls, the slip noose (the way rope halters are always made) tightens, and pinches his nose, and then he will struggle for life, until, perchance, he throws himself. But this is not the worst. *A horse that has once pulled on his halter can never be as well broken as one that has never pulled at all.*

Before anything more is attempted with the colt, some of the characteristics of his nature must be noticed, that his motions may be better understood. Every one that has ever paid any attention to the horse, has noticed his natural inclination to smell everything which to him looks new and frightful. This is their strange mode of examining everything. And when they are frightened at anything, though they look at it sharply, they seem to have no confidence in their eyesight alone, but must touch it with their nose before they are entirely satisfied; and, as soon as they have done that, all seems right.

If you want to satisfy yourself of this characteristic of the horse, and to learn something of importance concerning the peculiarities of his nature, etc., turn him into the barn yard, or a large stable will do, and then gather up something that you know will frighten him—a red blanket, buffalo robe, or something of that kind. Hold it up so that he can see it, he will stick up his head and snort. Then throw it down somewhere in the centre of the lot or barn, and walk off to one side. Watch his motions, and study his nature. If he is frightened at the object, he will not rest until he has touched it with his nose. He will begin to walk around the robe and snort, all the time getting a little closer, until he finally gets within reach of it. He will then very cautiously stretch out his neck as far as he can reach, merely touching it with his nose, as though he thought it was ready to fly at him. But after he has repeated these touches for a few times, for the first time (though he has been looking at it all the while) he seems to have an idea of what it is. When he has found, by the sense of feeling, that it is nothing that will do him any harm, he is ready to play with it. If you watch him closely, you will see him take hold of it with his teeth, and raise it up, and pull at it; and in a few minutes you can see that he has not that same wild look about his eye, but that he stands like a horse biting at some familiar stump.

STRAP FOR THE RIGHT FORE LEG. (See pages 366-370.)

Yet the horse is never so well satisfied when he is about anything that has frightened him, as when he is standing with his nose to it; and in nine cases out of ten, you will see some of that same wild look about him again, as he turns to walk from it. You will, probably, see him looking back very suspiciously as he walks away, as though he thought it might come after him yet. In all probability he will have to go back and make another examination before he is satisfied; but he will familiarize himself with it, and if he should run in that field for a few days, the robe that frightened him so much at first will be no more to him than a familiar stump.

It might very naturally be supposed from the fact of the horse's applying his nose to everything new to him, that he always does so for the purpose of smelling these objects; but it is as much or more for the purpose of feeling, and he makes use of his nose, or muzzle (as it is sometimes called), as we would of our hands; because it is the only organ by which he can touch or feel anything with much susceptibility.

He invariably makes use of the four senses—SEEING, HEARING, SMELLING, and FEELING—in all of his examinations, of which the sense of feeling is, perhaps, the most important. In the experiment with the robe, his gradual approach and final touch with his nose was as much for the purpose of feeling as anything else, his sense of smell being so keen that it would not be necessary for him to touch his nose against anything in order to get the proper scent; for it is said that a horse can smell a man at a distance of a mile. Besides, if the scent of the robe was all that was necessary, he could get that several rods off; whereas, we know from experience, that if a horse sees and smells a robe a short distance from him, he is very much frightened (unless he is used to it) until he touches or feels it with his nose; which is a positive proof that feeling is the controlling sense in this case.

It is a prevalent opinion among horsemen generally that the sense of smell is the governing sense of the horse; and with that view many receipts of strong-smelling oils, etc., have been concocted in order to tame him. All of these as far as the scent goes, have no effect whatever in taming him, or conveying any idea to his mind; though the acts that accompany these efforts—handling him, touching him about the nose and head, and patting him, as you are directed to do, after administering the articles, may have a very great effect, which is mistaken for the effect of the ingredients used.

APPROACHING A COLT.—In order to take horses as we find them, of all kinds, and to train them to our liking, we should always take with us, when we go into a stable to train a colt, a long switch whip (whalebone buggy whips are the best) with a good silk cracker, so as to cut keenly and make a sharp report.

STRAP FOR THE OFF FORE LEG. (See p. 370.) This, if handled with dexterity, and rightly applied, accompanied with a sharp, fierce word, will be sufficient to enliven the spirits of any horse. With this whip in your right hand, with the lash pointing backward, enter the stable alone. It is a great disadvantage in training a horse to have any one in the stable with you; you should be entirely alone, so as to have nothing but yourself to attract his attention. If he is wild, you will soon see him on the opposite side of the stable from you; and now is the time to use a little judgment.

Accordingly, when you have entered the stable, stand still, and let your horse look at you a minute or two, and as soon as he is settled in one place, approach him slowly, with both arms stationary, your right hanging by your side, holding the whip as directed, and the left bent at the elbow, with your hand projecting. As you approach him, go not too much toward his head or croup, so as not to make him move either forward or backward, thus keeping your horse stationary; if he does move a little either forward or backward, step a little to the right or left very cautiously; this will keep him in one place. As you get very near him, draw a little to his shoulder, and stop a few seconds. If you are in his reach he will turn his head and smell your hand, not that he has any preference for your hand, but because that is projecting, and is the nearest portion of your body to the horse. This all colts will do, and they will smell your naked hand just as quickly as they will of anything that you can put in it.

As soon as he touches your hand with his nose, caress him as before directed, always using a very light, soft hand, merely touching the horse, always rubbing the way the hair lies, so that your hand will pass along as smoothly as possible. As you stand by his side, you may find it more convenient to rub his neck or the side of his head, which will answer the same purpose as rubbing his forehead. Favor every inclination of the horse to smell or touch you with his nose. *Always follow each touch or communication of this kind with the most tender and affectionate caresses, accompanied with a kind look, and pleasant word of some sort,* such as, "Ho! my little boy—ho! my little boy!" "Pretty boy!" "Nice lady!" or something of that kind, constantly repeating the same words, with the same kind, steady tone of voice; for the horse soon learns to read the expression of the face and voice, and will know as well when fear, love, or anger prevails, as you know your own feelings; two of which, fear and anger, a good horseman should never feel.

If your horse, instead of being wild, seems to be of a stubborn or *mulish* disposition; if he lays back his ears as you approach him, or turns his heels to kick you, he has not that regard or fear of man that he should have, to enable you to handle him quickly and easily; and it might be well to give him a few sharp cuts with the whip, about the legs, pretty close to the body. It will crack keenly as it plies around his legs, and the crack of the whip will affect him as much as the stroke; besides, one sharp cut about his legs will affect him more than two or three over his back, the skin on the inner part of his legs or about his flank being thinner, more tender, than on his back. Do not whip him much—just enough to frighten him; it is not because we want to hurt the horse that we whip him—we only do it to frighten vice and stubbornness out of him. Whatever you do, do quickly, sharply, and with a good deal of fire, but always without anger. If you are going to frighten him at all, you must do it at once. Never go into a pitched battle with your horse, and whip him until he is mad and will fight you; it would be better not to touch him at all, for you will establish, instead of fear and respect, feelings of resentment, hatred, and ill-will. If you can succeed in frightening him, you can whip him without making him mad; for fear and anger never exist together in the horse, and as soon as one is visible, the other disappears. After you have frightened him, so that he will stand up straight and pay some attention to you, approach him again, and caress him a good deal more than you whipped him; then you will excite the two controlling passions of his nature, love and fear, and as soon as he learns what you require, he will obey quickly.

How to Halter and Lead a Colt.—As soon as you have tamed the colt a little, take the halter in your left hand, and approach him as before, and on the same side that you have tamed him. If he is very timid about your approaching closely to him, you can get up to him quicker by making the whip a part of your arm, and reaching out very gently with the butt end of it, rubbing him lightly on the neck, all the time getting a little closer, shortening the whip by taking it up in your hand, until you finally get close enough to put your hands on him. If he is inclined to hold his head from you, put the end of the halter-strap around his neck, drop your whip, and draw very gently; he will let his neck give, and you can pull his head to you. Then take hold of that part of the halter which buckles over the top of his head, and pass the long side, or that part which goes into the buckle, under his neck, grasping it on the opposite side with your right hand, letting the first strap loose —the halter will be sufficient to hold his head to you. Lower the halter a little, just enough to get his nose into that part which goes around it; then raise it somewhat, and fasten the top buckle, and you will have it all right. The first time you halter a colt you should stand on the left side, pretty well back to his shoulder, only taking hold of that part of the halter that goes around his neck; then with your two hands about his neck you can hold his head to you, and raise the halter on it without making him dodge by putting your hands about his nose. You should have a long rope or strap ready, and as soon as you have the halter on, attach this to it, so that you can let him walk the length of the stable without letting go of the strap, or without making him pull on the halter, for if you only let him feel the weight of your hand on the halter, and give him rope when he runs from you, he will never rear, pull, or throw himself, yet you will be holding him all the time, and doing more toward gentling him than if you had the power to snub him right up, and hold him to one spot; because he does not know anything about his strength, and if you don't do anything to make him pull, he will never know that he can. In a few minutes you can begin to control him with the halter; then shorten the distance between yourself and the horse by taking up the strap in your hand.

As soon as he will allow you to hold him by a tolerably short strap, and to step up to him without flying back. You can begin to give him some idea about leading. But to do this, do not go before and attempt to pull him after you, but commence by pulling him very quietly to one side. He has nothing to brace either side of his neck, and will soon yield to a steady, gradual pull of the halter; and as soon as you have pulled him a step or two to one side, step up to him and caress him, and then pull him again, repeating this operation until you can pull him around in every direction, and walk about the stable with him, which you can do in a few minutes, for he will soon think when you have made him step to the right or left a few times, that he is compelled to follow the pull of the halter, not knowing that he has the power to resist your pulling; besides, you have handled him so gently that he is not afraid of you, and you always caress him when he comes up to you, and he likes that, and would just as lief follow you as not. After he has had a few lessons of that kind, if you turn him out in a field, he will come up to you every opportunity he gets.

TAMING THE HORSE. (See page 368.)

You should lead him about in the stall some time before you take him out, opening the door so that he can see out, leading him up to it and back again, and past it. See that there is nothing on the outside to make him jump when you take him out, and as you go out with him, try to make him go very slowly, catching hold of the halter close to the jaw with your left hand, while the right is resting on the top of the neck, holding to his mane. After you are out with him a little while, you can lead him about as you please.

Don't let any second person come up to you when you first take him out; a stranger taking hold of the halter would frighten him, and make him run. There should not even be any one standing near him, to attract his attention or scare him. If you are alone, and manage him rightly, it will not require any more force to lead or hold him than it would to manage a broken horse.

How to Tie up a Colt.—If you want to tie up your colt, put him in a tolerably wide stall, which should not be too long, and should be connected by a bar or something of that kind to the partition behind it; so that, after the colt is in he cannot go far enough back to take a straight, backward pull on the halter; then by tying him in the centre of the stall, it would be impossible for him to pull on the halter, the partition behind preventing him from going back, and the halter in the centre checking him every time he turns to the right or left. In a stall of this kind you can break any horse to stand tied with a light strap, anywhere, without his ever knowing anything about pulling. For if you have broken your horse to lead, and have taught him the use of the halter (which you should always do before you hitch him to anything), you can hitch him in any kind of a stall, and if you give him something to eat to keep him up to his place for a few minutes at first, there is not one colt in fifty that will pull on his halter.

How to Tame a Horse.—Take up one fore-foot and bend his knee till his hoof is bottom upward, and nearly touching his body; then slip a loop over his knee, and up until it comes above the pastern-joint, to keep it up, being careful to draw the loop together between the hoof and pastern-joint with a second strap of some kind to prevent the loop from slipping down and coming off. This will leave the horse standing on three legs; you can now handle him as you wish, for it is utterly impossible for him to kick in this position. There is something in this operation of taking up one foot, that conquers a horse quicker and better than anything else you can do to him. There is no process in the world equal to it to break a kicking horse, as there is a principle of this kind in his nature that by conquering one member, you conquer, to a great extent, the whole horse.

This will conquer him better than anything you could do, and without any possible danger of hurting himself or you either, for you can tie up his foot and sit down and look at him until he gives up. When you find that he is conquered, go to him, let down his foot, rub his leg with your hand, caress him, and let him rest a little; then put it up again. Repeat this a few times, always putting up the same foot, and he will soon learn to travel on three legs, so that you can drive him some distance. As soon as he gets a little used to this way of traveling, put on your harness, and hitch him to a sulky. If he is the worst kicking horse that ever raised a foot, you need not be fearful of his doing any damage while he has one foot up, for he cannot kick, neither can he run fast enough to do any harm. And if he is the wildest horse that ever had harness on, and has run away every time he has been hitched, you can now hitch him in a sulky, and drive him as you please. If he wants to run, you can let him have the lines, and the whip too, with perfect safety, for he can go but a slow gait on three legs, and will soon be tired, and willing to stop; only hold him enough to guide him in the right direction, and he will soon be tired and willing to stop at the word. Thus you will effectually cure him at once of any further notion of running off. Kicking horses have always been the dread of everybody; but by this new method you can harness them to a rattling sulky, plough, wagon, or anything else in its worst shape. They may be frightened at first, but cannot kick, or do anything to hurt themselves, and will soon find that you do not intend to hurt them, and then they will not care anything more about it. You can then let down the leg and drive along gently without any further trouble. By this new process a bad kicking horse can be taught to go gentle in harness in a few hours' time.

How to Make a Horse lie down.—To make a horse lie down, bend his left fore-leg and slip a loop over it, so that he cannot get it down. Then put a surcingle round his body, and fasten one end of a long strap around the other fore-leg, just above the hoof. Place the other end under the before-described surcingle, so as to keep the strap in the right direction; take a short hold of it with your right hand; stand on the left side of the horse, grasp the bit in your left hand, pull steadily on the strap with your right; bear against his shoulder till you cause him to move. As soon as he lifts his weight, your pulling will raise the other foot, and he will have to come on his knees.

Keep the strap tight in your hand, so that he cannot straighten his leg if he rises up. Hold him in this position, and turn his head toward you; bear against his side with your shoulder, not hard, but with a steady, equal pressure, and in about ten minutes he will lie down. As soon as he lies down, he will be completely conquered, and you can handle him as you please. Take off the straps, and straighten out his legs; rub him lightly about the face and neck with your hand the way the hair lies; handle all his legs, and after he has lain ten or twenty minutes, let him get up again. After resting him a short time, make him lie down as before. Repeat the operation three or four times, which will be sufficient for one lesson. Give him two lessons a day, and when you have reached four lessons, he will lie down by taking hold of one foot. As soon as he is well broken to lie down in this way, tap him on the opposite leg with a stick when you take hold of his foot, and in a few days he will lie down from the mere motion of the stick.

TEACHING THE HORSE TO LIE DOWN

To Accustom a Horse to Strange Sounds and Sights.—It is an excellent practice to accustom all horses to strange sounds and sights, and of very great importance to young horses which are to be ridden or driven in large towns, or used as chargers. Although some horses are very much more timid and nervous than others, the very worst can be very much improved by acting on the first principles laid down in the introduction to this article—that is, by proving that the strange sights and sounds will do them no harm.

When a railway is first opened, the sheep, the cattle, and especially the horses, grazing in the neighboring fields, are terribly alarmed at the sight of the swift, dark, moving trains, and the terrible snorting and hissing of the steam engines. They start away—they gallop in circles—and when they stop, gaze with head and tail erect, until the monsters have disappeared. But from day to day the live stock become more accustomed to the sight and sound of the steam horse, and after a while they do not even cease grazing when the train passes. They have learned that it will do them no harm. The same result may be observed with respect to young horses when first they are brought to a large town, and have to meet great loads of hay, omnibuses crowded with passengers, and other strange or noisy objects; if judiciously treated, not flogged and ill-used, they lose their fears without losing their high courage.

To accustom a Horse to a Drum.—Place it near him on the ground, and without forcing him, induce him to smell it again and again, until he is thoroughly accustomed to it. Then lift it up, and slowly place it on the side of his neck, where he can see it, and tap it gently with a stick or your finger. If he starts, pause, and let him carefully examine it. Then commence again, gradually moving it backward until it rests upon his withers, by degrees playing louder and louder, pausing always when he seems alarmed, to let him look at it and smell, if needful. In a very few minutes you may play with all your force, without his taking any notice. When this practice has been repeated a few times, your horse, however spirited, will rest his nose unmoved on the big drum, while the most thundering piece is played.

To teach a Horse to bear an Umbrella.—Go through the same cautious forms, let him see it, and smell it, open it by degrees, gain your point inch by inch, passing it always from his eyes to his neck, and from his neck to his back and tail; and so with a riding-habit; in half an hour any horse may be taught that it will not hurt him, and then the difficulty is over.

To fire off a Horse's back.—Begin with caps, and, by degrees, as with the drum. Instead of lengthening the reins, stretch the bridle hand to the front, and raise it for the carbine to rest on, with the muzzle clear of the horse's head, a little to one side. Lean the body forward without rising in the stirrups. *Avoid interfering with the horse's mouth, or exciting his fears by suddenly closing your legs either before or after firing—be quiet yourself, and your horse will be quiet.* The colt can learn to bear a rider on his bare back during his first lessons, when prostrate and powerless, fast bound by straps. The surcingle has accustomed him to girths, he leads well, and has learned that when the right rein is pulled he must go to the right, and when the left rein to the left. You may now teach him to bear the BIT and the SADDLE, if you have not placed it upon his back while on the ground.

HOW TO ACCUSTOM A HORSE TO A BIT.—You should use a large, smooth, snaffle bit, so as not to hurt his mouth, with a bar to each side, to prevent the bit from pulling through either way. This you should attach to the head-stall of your bridle, and put it on your colt without any reins to it, and let him run loose in a large stable or shed some time, until he becomes a little used to the bit, and will bear it without trying to get it out of his mouth. It would be well, if convenient, to repeat this several times, before you do anything more with the colt; as soon as he will bear the bit, attach a single rein to it. You should also have a halter on your colt, or a bridle made after the fashion of a halter, with a strap to it, so that you can hold or lead him about without pulling at the bit much. He is now ready for the saddle.

STRUGGLES OF THE VICIOUS HORSE AGAINST LYING DOWN.

THE PROPER WAY TO BIT A COLT.—Farmers often put bitting harness on a colt the first thing they do to him, buckling up the bitting as tight as they can draw it, to make him carry his head high, and then turn him out in a field to run half a day at a time. This is one of the worst punishments that could be inflicted on the colt, and is very injurious to a young horse that has been used to running in pasture with his head down. Colts have been so seriously injured in this way that they have never recovered.

A horse should be well accustomed to the bit before you put on the bitting harness, and when you first bit him you should only rein his head up to that point where he naturally holds it, let that be high or low; he will soon learn that he cannot lower his head, and that raising it a little will loosen the bit in his mouth. This will give him the idea of raising his head to loosen the bit, and then you can draw the bitting a little tighter every time you put it on, and he will still raise his head to loosen it; by this means you will gradually get his head and neck in the position you want him to carry them, and give him a nice and graceful carriage without hurting him, making him mad, or causing his mouth to get sore.

If you put the bitting on very tight the first time, he cannot raise his head enough to loosen it, but will bear on it all the time, and paw, sweat, and throw himself. Many horses have been killed by falling backward with the bitting on; their heads being drawn up strike the ground with the whole weight of the body. Horses that have their heads drawn up tightly should not have the bitting on more than fifteen or twenty minutes at a time.

How to Saddle a Colt.—The first thing will be to tie each stirrup-strap into a loose knot to make them short, and prevent the stirrups from flying about and hitting him. Then double up the skirts and take the saddle under your right arm, so as not to frighten him with it as you approach. When you get to him rub him gently a few times with your hand, and then raise the saddle very slowly, until he can see it, and smell and feel it with his nose. Then let the skirt loose, and rub it very gently against his neck the way the hair lies, letting him hear the rattle of the skirts as he feels them against him; each time getting a little further backward, and finally slipping it over his shoulders on his back. Shake it a little with your hand, and in less than five minutes you can rattle it about over his back as much as you please, and pull it off and throw it on again, without his paying much attention to it.

As soon as you have accustomed him to the saddle, fasten the girth. Be careful how you do this. It often frightens the colt when he feels the girth binding him, and making the saddle fit tight on his back. You should bring up the girth very gently, and not draw it too tight at first, just enough to hold the saddle on. Move him a little, and then girth it as tight as you choose, and he will not mind it.

You should see that the pad of your saddle is all right before you put it on, and that there is nothing to make it hurt him, or feel unpleasant to his back. It should not have any loose straps on the back part of it, to flap about and scare him. After you have saddled him in this way, take a switch in your right hand to tap him up with, and walk about in the stable a few times with your right arm over your saddle, taking hold of the reins on each side of his neck with your right and left hands, thus marching him about in the stable until you teach him the use of the bridle and can turn him about in any direction, and stop him by a gentle pull of the rein. Always caress him, and loose the reins a little every time you stop him.

You should always be alone, and have your colt in some light stable or shed, the first time you ride him; the loft should be high, so that you can sit on his back without endangering your head. You can teach him more in two hour's time in a stable of this kind, than you could in two weeks in the common way of breaking colts, out in an open place. If you follow my course of treatment, you need not run any risk, or have any trouble in riding the worst kind of horse. You take him a step at a time, until you get up a mutual confidence and trust between yourself and horse. First teach him to lead and stand hitched; next acquaint him with the saddle, and the use of the bit; and then all that remains is to get on him without scaring him, and you can ride him as well as any horse.

How to Mount the Colt.—First gentle him well on both sides, about the saddle, and all over until he will stand still without holding, and is not afraid to see you anywhere about him. As soon as you have him thus gentled, get a small block, about one foot or eighteen inches in height, and set it down by the side of him, about where you want to stand to mount him; step up on this, raising yourself very gently; horses notice every change of position very closely, and if you were to step up suddenly on the block, it would be very apt to scare him; but by raising yourself gradually on it, he will see you without being frightened, in a position very nearly the same as when you are on his back.

As soon as he will bear this without alarm, untie the stirrup-strap next to you, and put your left foot into the stirrup, and stand square over it, holding your knee against the horse, and your toes out, so as not to touch him under the shoulder with the toe of your boot. Place your right hand on the front of the saddle, and on the opposite side of you, taking hold of a portion of the mane and the reins, as they hang loosely over his neck, with your left hand; then gradually bear your weight on the stirrup, and on your right hand, until the horse feels your whole weight on the saddle; repeat this several times, each time raising yourself a little higher from the block, until he will allow you to raise your leg over his croup, and place yourself in the saddle.

There are three great advantages in having a block from which to mount. First, a sudden change of position is very apt to frighten a young horse who has never been handled; he will allow you to walk up to him, and stand by his side without scaring at you, because you have gentled him to that position; but if you get down on your hands and knees and crawl toward him, he will be very much frightened; and upon the same principle, he

SUBMISSION OF THE HORSE.

would be frightened at your new position if you had the power to hold yourself over his back without touching him. The first great advantage of the block, then, is to gradually gentle him to that new position in which he will see you when you ride him.

Secondly, by the process of leaning your weight in the stirrup, and on your hand, you can gradually accustom him to your weight, so as not to frighten him by having him feel it all at once. And, in the third place, the block elevates you so that you will not have to make a spring in order to get upon the horse's back, but from it you can gradually raise yourself into the saddle. When you take these precautions, there is no horse so wild but that you can mount him without making him jump. When mounting, your horse should always stand without being held. A horse is never well broken when he has to be held with a tight rein when mounting; and a colt is never so safe to mount as when you see that assurance of confidence, and absence of fear, which cause him to stand without holding.

An improved plan of mounting is to pass the palm of the right hand on the off-side of the saddle, and as you rise lean your weight on it; by this means you can mount with the girth loose, or without any girth at all.

How to Ride a Colt.—When you want him to start do not touch him on the side with your heel, or do anything to frighten him and make him jump. But speak to him kindly, and if he does not start pull him a little to the left until he starts, and then let him walk off slowly with the reins loose. Walk him around in the stable a few times until he gets used to the bit, and you can turn him about in every direction and stop him as you please. It would be well to get on and off a good many times until he gets perfectly used to it before you take him out of the stable.

After you have trained him in this way, which should not take you more than one or two hours, you can ride him anywhere you choose without ever having him jump or make any effort to throw you.

When you first take him out of the stable be very gentle with him, as he will feel a little more at liberty to jump or run, and be a little easier frightened than he was while in the stable. But after handling him so much in the stable he will be pretty well broken, and you will be able to manage him without trouble or danger.

When you first mount him take a little the shortest hold on the left rein, so that if anything frightens him you can prevent him from jumping by pulling his head round to you. This operation of pulling a horse's head round against his side will prevent any horse from jumping ahead, rearing up, or running away. If he is stubborn and will not go, you can make him move by pulling his head round to one side, when whipping would have no effect. And turning him round a few times will make him dizzy, and then by letting him have his head straight, and giving him a little touch with the whip, he will go along without any trouble.

Never use martingales on a colt when you first ride him; every movement of the hand should go right to the bits in the direction in which it is applied to the reins, without a martingale to change the direction of the force applied. You can guide the colt much better without it, and teach him the use of the bit in much less time. Besides, martingales would prevent you from pulling his head round if he should try to jump.

After your colt has been ridden until he is gentle and well accustomed to the bit, you may find it an advantage, if he carries his head too high or his nose too far out, to put martingales on him.

You should be careful not to ride your colt so far at first as to heat, worry, or tire him. Get off as soon as you see that he is a little fatigued; gentle him, and let him rest; this will make him kind to you, and prevent him from getting stubborn or mad.

To Break a Horse to Harness.—Take him in a light stable, as you did to ride him; take the harness, and go through the same process that you did with the saddle, until you get him.

familiar with it, so that you can put it on him, and rattle it about without his caring for it. As soon as he will bear this, put on the lines, caress him as you draw them over him, and drive him about in the stable till he will bear them over his hips. The lines are a great aggravation to some colts, and often frighten them as much as if you were to raise a whip over them. As soon as he is familiar with the harness and lines, take him out and put him by the side of a gentle horse. Always use a bridle without blinkers when you are breaking a horse to harness.

Lead him to and around a light gig or phaeton; let him look at it, touch it with his nose, and stand by it till he does not care for it: then pull the shafts a little to the left, and stand your horse in front of the off-wheel. Let some one stand on the right side of the horse, and hold him by the bit, while you stand on the left side, facing the sulky. This will keep him straight. Run your left hand back, and let it rest on his hip, and lay hold of the shafts with your right, bringing them up very gently to the left hand, which still remains stationary.

BREAKING THE HORSE TO HARNESS.

Do not let anything but your arm touch his back, and as soon as you have the shafts square over him, let the person on the opposite side take hold of one of them, and lower them very gently to the shaft-bearers. Be very slow and deliberate about hitching; the longer time you take the better, as a general thing. When you have the shafts placed, shake them slightly, so that he will feel them against each side. As soon as he will bear them without scaring, fasten your braces, etc., and start him along very slowly. Let one man lead the horse, to keep him gentle, while the other gradually works back with the lines till he can get behind and drive him. After you have driven him in this way a short distance, you can get into the sulky, and all will go right. It is very important to have your horse go gently when you first hitch him. After you have walked him awhile, there is not half so much danger of his scaring. Men do very wrong to jump up behind a horse to drive him as soon as they have him hitched. There are too many things for him to comprehend all at once. The shafts, the lines, the harness, and the rattling of the sulky, all tend to scare him, and he must be made familiar with them by degrees. If your horse is very wild, one foot had better be put up the first time you drive him. With the leg strapped up, the lighter the gig the better, and four wheels are better than two.

WARRANTY.

In the purchase of a horse the buyer should take with the receipt what is termed in law a warranty. The best way of expressing it is in this form:

Philadelphia, August 1, 18—.

Received of William Ingalls three hundred dollars, for a black mare, warranted only five years old, sound, free from vice, and quiet to ride and drive.

$300.   EDWARD RIDDLE.

A receipt, which includes simply the word "warranted," extends merely to soundness. "Warranted sound," has no greater extent; the age, freedom from vice, and quietness to ride and drive should all be especially named. This warranty embraces every cause of unsoundness that can be detected, or that is inherent in the constitution of the animal at the time of sale, as well as every vicious habit which he has previously shown. In order to establish a breach of the warranty, and then be enabled to return the horse or recover the price paid, the purchaser must prove that it was unsound or viciously disposed at the time of sale. In case of cough, the horse must have been heard to cough previously to the purchase, or as he was led home, or as soon as he had entered the stable of the purchaser. Coughing, even on the following morning, will not be sufficient; for it is possible that he might have caught cold by a change of stabling. If he is lame, it must be proved to arise from a cause that could not have occurred after he was in the purchaser's possession. No price will imply a warranty, or be deemed equivalent to one; the warranty must be expressly stated.

A fraud in the seller must be proved, in order that the buyer may be enabled to return the horse or maintain an action for the price. The warranty should be given at the time of sale. A warranty or a promise to warrant the horse, given at any period previous to the sale, is of no effect; for the horse is a very perishable commodity, and his constitution and his usefulness may undergo a considerable change in a few days. A warranty after the sale is also of no effect, as it is given without any legal consideration. In order to complete the purchase, there must be a transfer of the animal, or a written memorandum of agreement, or the payment of some sum, however small, as earnest-money. No verbal promise to buy or sell is binding without one of these accompaniments; and the moment either of them is effected, the legal transfer of property, or its delivery, is made, and whatever may happen to the horse, the seller retains, or is entitled to, the money. If the purchaser exercises any act of ownership—as by using the animal without leave of the seller, or by having any operation performed upon him, or medicines given to him—he makes him his own.

If the horse should afterward be discovered to have been unsound at the time of warranty and sale, the buyer may return him. Although not legally compelled to give notice to the seller of the discovered unsoundness, it is best that such notice should be given. The animal should then be tendered at the house or stable of the seller. If he refuses to receive the animal, humanity dictates that he should be sent to a livery stable, in preference to tying him up in the street; an action can be maintained, after the horse has been tendered, for the necessary expenses of keeping him as well as for the price paid. The keep, however, can be recovered only for the time that necessarily intervened between the tender and the determination of the action. It is not legally necessary to return the animal as soon as the unsoundness is discovered. The animal may be kept for a reasonable time afterward, and even proper medical means may be resorted to for the removal of the unsoundness; but courtesy, and indeed justice, will require that the notice should be given as soon as possible. Although it is laid down, upon the authority of an eminent English judge, that "no length of time elapsed after the sale, will alter the nature of a contract originally false," yet there are recorded cases in which the buyer was prevented from maintaining his action, because he did not give notice of the unsoundness within a reasonable time after its discovery. What will constitute this reasonable time, depends upon many circumstances. It was formerly supposed that the buyer had no right to have the horse medically treated, and that he would vitiate the warranty by so doing. The question, however, in such a case would be, whether the animal was injured, or his value lessened, by such treatment. It may be remarked that it is generally most prudent to refrain from all medical treatment, since the means adopted, no matter how skillfully used, may have an unfortunate effect, or what is done may be misrepresented by ignorant or interested observers.

When a horse is returned, and an action brought for the price, it is indispensable that in every respect, except the alleged unsoundness, the animal should be as perfect and valuable as when he was bought.

The purchaser may, possibly, like the horse, notwithstanding his discovered defect; in which case he may retain him and bring an action for the depreciation in value on account of the unsoundness. Few, however, will do this, because the retaining of the animal will give rise to a suspicion that the defect is of no great consequence, and consequently will occasion much cavil about the amount of damages; the suit terminating, probably, in the recovery of slight, if any, damages.

Where there is no warranty, an action may be brought on the ground of fraud; but as this is very difficult to be maintained, few persons will hazard it. It will in such a case, be necessary to prove that the seller knew the defect, and that the buyer was imposed upon by his false representations; and that, too, under circumstances in which a person of ordinary carefulness and circumspection might have been imposed upon. If the defect was palpably evident, the purchaser has no remedy, for he should have exercised more caution; but if a warranty was given, it covers every unsoundness, evident or concealed. Although a person should ignorantly or carelessly buy a blind horse, warranted sound, he may return it—the warranty is his protection, and prevents him from examining the horse as closely as he otherwise would have done; but if he buys a blind horse, supposing him to be sound, and without a warranty, he is without any remedy. The law supposes every one to exercise common circumspection and common sense.

A person should have a more thorough knowledge of horses than most possess, together with perfect confidence in the seller, who ventures to buy a horse without a warranty. If a person buy a horse warranted sound, and discovering no defect in him, sells him again, relying upon his warranty, and the unsoundness is discovered by the second purchaser, and the horse returned to the first buyer, or an action commenced against him, the latter has his claim upon the first seller, and may demand of him not only the price of the horse, or the difference in value, but all expenses which may necessarily have been incurred.

Exchanges, whether of one horse absolutely for another, or where a sum of money is paid in addition by one of the parties, stand upon precisely the same ground as simple sales. If there is a warranty upon either side, and that is broken, the exchange is vitiated; if there is no warranty, deceit must be proved.

THE END.

www.ingramcontent.com/pod-product-compliance
Lightning Source LLC
Chambersburg PA
CBHW030358230426
43664CB00007BB/651